Bioinked Boundaries

Pratap Devarapalli

Bioinked Boundaries

Patenting 3D Bioprinted Tissues, Organs and Bioinks: An US, European and Australian Patent Law Perspective

Springer

Pratap Devarapalli
School of Law
University of Queensland
Brisbane, QLD, Australia

ARC Centre of Excellence in Synthetic Biology
Brisbane, Australia

ARC Centre of Excellence in Quantum Biotechnology
Brisbane, Australia

Centre of Law and Genetics, Faculty of Law
University of Tasmania
Hobart, Australia

ISBN 978-981-96-2590-1 ISBN 978-981-96-2591-8 (eBook)
https://doi.org/10.1007/978-981-96-2591-8

This Springer imprint is published by the registered company Springer Nature Singapore Pte Ltd.
The registered company address is: 152 Beach Road, #21-01/04 Gateway East, Singapore 189721, Singapore

Foreword

Patent law has always been closely tied to the innovations that it seeks to regulate. From the organic chemicals of the nineteenth century to computer-related inventions in the later part of the twentieth century, each new wave of innovation has pushed the boundaries of the law. Science and technology have not only consistently provided the law with potential new candidates for protection, but they have also played an important role in helping the law deal with and accommodate that new subject matter. *Bioinked Boundaries* addresses one of the latest of these challenges: the patentability of bioprinted tissues and bioinks.

3D bioprinting, which sees the life sciences, computational technologies, and medical innovations converging in novel ways, creates a number of problems for the law; primarily because of the way bioprinted innovations blend the natural and synthetic. Focusing on patentable subject matter, *Bioinked Boundaries* provides a two-phase comparative analysis that combines doctrinal and empirical analysis to explore the fate of bioprinted tissues and bioinks in the United States, Europe, and Australia. The doctrinal analysis reveals the nuanced and often inconsistent application of patent laws across these three jurisdictions. It also highlights the difficulties that the courts face in determining the level of human intervention required for bioprinted tissues to meet the subject matter threshold and in distinguishing these innovations from their natural counterparts. The doctrinal analysis is complemented by an empirical analysis that provides data-driven insights into the practices of patent examiners and how these impact bioprinting's path to patentability.

As well as providing a comparative study of the fate of bioprinted tissues as patentable subject matter, *Bioinked Boundaries* also considers patent law's adaptability to change. *Bioinked Boundaries* does not, however, merely identify problems; it also offers patent applicants with the tools to guide them through the labyrinthine process of securing patents for bioprinting technologies in the United States, Europe, and Australia. *Bioinked Boundaries* not only provide an important resource for legal practitioners, policymakers, and innovators, but they also challenge us to rethink the boundaries of patent law and to consider how we can best support the technologies that have the potential to transform human lives.

Prof. Brad Sherman
Laureate Fellow
The University of Queensland
Brisbane, Australia

Acknowledgements

First and foremost, I extend my deepest gratitude to my parents for their unwavering love and support throughout my life. Their encouragement has given me the strength to pursue my dreams and reach for the stars. I am also profoundly grateful to my wife, Seema, whose steadfast support has been invaluable during this journey. Her patience, understanding, and willingness to shoulder many family responsibilities have allowed me to focus fully on my work. To our little prince, Aaron, who arrived early in this journey: thank you for your patience, even when Dad was engrossed in his research. A big hug to you, my son.

I also wish to thank Distinguished Prof. Dianne Nicol and Prof. Jane Nielsen at the University of Tasmania for their invaluable guidance and support during the research that forms a significant part of this book. I am further indebted to Australian Laureate Fellow Professor Brad Sherman and Associate Professor Allison Fish for their insightful feedback and constructive critiques, which have enriched this work in countless ways.

The research conducted in this book was supported by three Australian Research Council (ARC) initiatives: the Discovery Project 'Reforming the Regulatory Environment for Innovative Health Technologies' (DP180101262) at the University of Tasmania's Centre for Law and Genetics; the ARC Centre of Excellence in Synthetic Biology (CE200100029); and the ARC Centre of Excellence in Quantum Biotechnology (CE230100021). I extend my sincere gratitude to each of these institutions for their vital support and dedication to advancing cutting-edge research and innovation.

Contents

Chapter 1
Introduction

How does the law evolve in response to technological revolutions? What happens when innovations outpace legal frameworks, challenging the definitions and boundaries that once seemed clear? These questions have haunted legal systems for centuries, from the invention of the steam engine to contemporary frontier life science research, ranging from synthetic biology through to quantum-driven biotechnologies. In this sense, every new technology has the potential to give rise to new questions that stretch the limits of existing legal systems, forcing courts and regulators to assess new scenarios and adapt existing legal systems. This tension is particularly apparent in the field of biotechnology, where breakthroughs are not only altering our ability to manipulate natural biological systems but are also altering industries as diverse as health and energy, as well as reshaping fundamental legal concepts. One such breakthrough, three-dimensional (3D) bioprinting, is at the forefront of this challenge for several reasons. Particularly, 3D bioprinting raises profound questions about the adequacy of current patent systems, as it compels us to reconsider whether, and in what ways, inventions transform naturally existing elements.

Intellectual property laws around the globe are designed to provide protection for intangible assets with patents intended as a legal mechanism capable of incentivising innovation across different fields of technology by providing limited term monopoly rights, thus preventing the 'free rider' problem.[1] It is believed that these limited term monopolies enable patentees to recoup investment in research and development.[2] The patent system also provides public access to new technologies through the disclosure requirement, which enables others to work outside the patent boundaries and to use the patented technology once the patent has expired.[3] Under this logic, patents play

[1] Ishac Diwan and Dani Rodrik, 'Patents, appropriate technology, and North–South trade' (1991) 30(1–2) *Journal of International economics* 27–47.

[2] Roberto Mazzoleni and Richard R. Nelson, 'The benefits and costs of strong patent protection: a contribution to the current debate' (1998) 27(3) *Research policy* 273–284.

[3] David Encaoua, Dominique Guellec and Catalina Martínez, 'Patent systems for encouraging innovation: Lessons from economic analysis' (2006) 35(9) *Research policy* 1423–1440.

P. Devarapalli, *Bioinked Boundaries*, https://doi.org/10.1007/978-981-96-2591-8_1

a crucial role in the entire technology life cycle. The economic benefits of the patent system are derived from its role in promoting innovation, encouraging investment, economic growth, knowledge sharing through disclosure, and the efficient use of resources.[4] Specifically, patenting inventions in a competitive research space are used as levers to attract financial investment. However, it is also important to note that patent rights do not provide any guarantee on commercialisation of a patented invention as this depends on multiple factors such as industrial applicability and commercial capability.[5]

Internationally, the patent system is governed by the Patent Cooperation Treaty (PCT) and the Agreement on Trade-related Aspects of Intellectual Property Rights (TRIPS), international treaties which set minimum standards for patentability.[6] These standards must be adopted by member states and integrated into jurisdiction-specific domestic law; however, different member states have some latitude in implementation of standards and interpretation of patentability criteria. As a result, there is some variation in patent law across various jurisdictions. As a result, a patentable invention must fall within the scope of the patentability criteria as set out in that specific jurisdiction. An invention will only be considered to be patentable if it satisfies the threshold requirement of patentable subject matter and if satisfies other patentability requirements[7] which include novelty, inventive step/non-obviousness, utility/usefulness/industrial applicability, and the disclosure requirements (enablement and written description).[8] The kinds of subject matter that fail to satisfy the patentable subject matter requirement include mathematical methods, scientific theories, discoveries, and, in some jurisdictions, plants and animals, as such subject matter has no technical effect and is therefore not an invention.[9] This question of patentable subject matter has been subject to intense judicial and scholarly scrutiny over the past two decades, particularly in relation to computer-implemented subject matter and subject matter from the nature.[10] It is for that reason that the primary inquiry of this book is the threshold question of what is patentable subject matter in relation to bioprinting innovation?

---

[4] Australian Law Reform Commission. *Genes and ingenuity: gene patenting and human health.* (Report 99, August 2004).

[5] Ismail Kamariah, Zaidi Wan Omar Wan and Abdul Majid Izaidin, 'The commercialisation process of patents by universities' (2011) 5(17) *African Journal of Business Management* 7198–7208.

[6] *Marrakesh Agreement Establishing the World Trade Organization*, opened for signature 15 April 1994, 1867 UNTS 3 (entered into force 1 January 1995) annex 1C (*Agreement on Trade-Related Aspects of Intellectual Property Rights*) ('*TRIPS*').

[7] May differ according to the jurisdiction.

[8] Giles S. Rich, 'Principles of Patentability' (1959) 28 *The George Washington Law Review* 393.

[9] *Patents Act 1990* (Cth) s 18(2).

[10] *D'Arcy v Myriad Genetics Inc* [2015] HCA 35; *Association for Molecular Pathology v Myriad Genetics Inc*, 133 US 2107 (2013); *Bilski v Kappos*, 561 US 593 (2010); *Mayo Collaborative Services v Prometheus Laboratories, Inc*, 566 US 66 (2012); *Alice Corporation Pty Ltd v CLS Bank International*, 134 US 2347 (2014); Brad Sherman, *Intangible Intangibles: Patent Law's Engagement with Dematerialised Subject Matter* (Cambridge University Press, 2024) vol 63.

The seeds of our contemporary international patent systems began to emerge approximately 500 years ago and, in the ensuing centuries, have been gradually changing. A significant driver of this change has been the adaptation required by the continual stream of legal challenges and questions that have arisen around new and emerging technologies.[11] At its core, the early patent systems were designed to promote innovation by granting inventors exclusive rights to their creations in exchange for public disclosure. The earliest patents, dating back to the fifteenth century, were granted for mechanical inventions—tools, machines, and processes that were easily defined and understood. These inventions were clear products of human ingenuity, and their creators were rewarded with legal monopolies allowing them to commercialise their innovations without fear of competition.

As technology advanced, however, the boundaries of invention became less clear. The industrial revolution brought with it a wave of new technologies, many of which relied on chemical processes and new forms of energy. Patent law had to evolve to accommodate these new inventions and courts began to wrestle with questions about what types of innovations deserved legal protection. The rise of chemistry in the nineteenth century, for example, introduced new challenges. Inventions like synthetic dyes and pharmaceuticals relied on natural materials, but they involved human intervention that transformed these materials into something new. Courts began to develop legal doctrines to address these challenges, expanding the scope of patentable subject matter to include chemical compositions and manufacturing processes that were not purely mechanical.[12]

One of the key domains of innovation that challenged patent law throughout the twentieth century was biotechnology. As advances in DNA research and techniques for gene manipulation unfolded, new possibilities emerged for innovation in agriculture, medicine, and industry. Biotechnology, involving the visualisation and manipulation of living organisms, raised significant policy, ethical, and legal questions around the ownership of life itself. This left courts and patent offices to grapple with whether, and to what extent, these new forms of innovation (many of which involved living organisms) could be eligible for patent protection. In 1980, the U.S. Supreme Court answered this question with its landmark decision in *Diamond v. Chakrabarty*, ruling that a genetically modified bacterium designed to consume oil spills was eligible for patent protection. The Court reasoned that the bacterium, though living, was the result of human intervention and therefore qualified as an invention under patent law.[13]

The decision of *Diamond v. Chakrabarty* marked a turning point in the evolution of patent law, setting the stage for the industrial expansion of biotechnology. In the years that followed, patent offices around the globe granted thousands of patents on

---

[11] Wyndham, Hulme E., 'History of the Patent System under the Prerogative and at Common Law' (1896) 12 *Law Quarterly Review* 141.

[12] *Le Roy v. Tatham* [1852]; Paul Lucier, 'Court and controversy: patenting science in the nineteenth century' (1996) 29(2) *The British Journal for the History of Science* 139–154.

[13] *Diamond v. Chakrabarty*, 447 U.S. 303, 309 [1980].

genetically modified organisms, gene sequences, proteins, and other biological materials. But as the scope of biotechnology patents expanded, so too did the controversy surrounding them. Critics argued that allowing patents on genes and living organisms commodified nature and restricted access to important medical treatments.[14] While this criticism has a minimal impact from a judicial perspective, the question of genes being patentable subject matter has been a significant judicial concern. The 2013 U.S. Supreme Court decision in *Association for Molecular Pathology v. Myriad Genetics*, for example, invalidated patents on isolated DNA sequences, ruling that naturally occurring genes could not be patented even if they had been isolated in a laboratory.[15]

The emergence of new technologies and scientific advancements such as biotechnology, computer-related inventions, and nanotechnology have raised many questions about the scope of patentable subject matter.[16] In the case of biotechnology and computer-related inventions, patent offices have introduced specific guidelines to assist patent examiners in assessing these complex technologies.[17] A body of case law has emerged in each of these respective areas that highlights the difficulty courts have had in determining whether innovations in these areas satisfy the fundamental thresholds of patentability. A gradual increase in innovation in different technology fields has created a need to expand the scope of patentable subject matter so that these novel technologies can be legislatively incentivised.[18]

It is against this backdrop of evolving legal doctrine and the ongoing ethical and social controversy around the technological manipulation of living organisms that the patentability of bioprinting must be understood. 3D bioprinting, a technology that allows for the creation of human tissues and, potentially, entire organs, stands at the cutting edge of innovation in the life sciences. Three-dimensional bioprinting involves the use of bioinks, composed of living stem cells, to print complex organic structures layer by layer, mimicking the architecture of biological tissues. The implications for industrial applications are varied and the potential financial impact and human benefit are staggering. For example, in medicine bioprinted tissues could revolutionise drug testing, eliminate the need for animal models, and aim to offer solutions to the global organ shortage. Bioprinted tissues, much like genetically modified organisms, involve both natural materials and human intervention. However, as with many biotechnological advancements, the legal questions that surround bioprinting

[14] Mark J. Hanson, 'Biotechnology and commodification within health care' (1999) 24(3) *The Journal of medicine and Philosophy* 267–287.

[15] *Association for Molecular Pathology v Myriad Genetics Inc*, 133 US 2107 (2013).

[16] John H. Barton, 'Adapting the intellectual property system to new technologies.' (1995) 10, no. 2–3 *International Journal of Technology Management* 151–172.

[17] UK Intellectual Property Office, 'Examination Guidelines for Patent Applications relating to Biotechnological Inventions in the Intellectual Property Office' 2010. https://www.gov.uk/government/publications/examining-patent-applications-for-biotechnological-inventions; Intellectual Property India, 'Guidelines for Examination of Computer Related Inventions (CRIs)' 2017.

[18] Michael, North. 'The US Expansion of Patentable Subject Matter: Creating a Competitive Advantage for Foreign Multinational Companies.' (2000) 18 *Boston University International Law Journal* 111.

are equally complex. Can a tissue printed from living cells be considered an invention? How much modification is required to transform a biological material into something that qualifies as patentable subject matter? And how should courts and patent offices balance the need to protect innovation with the need to ensure public access to important medical technologies? Are bioprinted tissues and bioinks patentable subject matter under current patent laws? If so, how should courts and patent offices draw the line between patentable inventions and unpatentable natural phenomena? Is the patentable subject matter requirement a hurdle to patenting bioprinting inventions, specifically bioinks and bioprinted tissues? And how do different jurisdictions approach these questions in light of their own patent laws?

3D bioprinting is a technology that represents the combination of the above-mentioned areas of technological innovation (i.e., life sciences, computation, and medicine). Specifically, the types of subject matter related to 3D bioprinted tissues and bioinks are complex, as they are composed of naturally available elements such as living stem cells in combination with natural and/or synthetic bioactive molecules[19] and other biomaterials that are 3D bioprinted in the form of a specific structure or scaffold. This book considers these complexities and examines whether, when, and to what extent 3D bioprinting innovation fit within the scope of patentable subject matter, or if the current narrowing of this legal concept threatens the patentability of these core technologies. If deemed patentable, these protections could incentivise further research and investment in bioprinting; conversely, restricting their patentability may hinder innovation by limiting the commercial incentives needed to advance this transformative field.

3D printing is a distinctive technology that has impacted everyday world.[20] 3D printing works on the same principle as a 2D printer but instead of words the machine can read a 3D design and lay down successive layers of liquid powder or sheet material to create almost any shape that can be designed on a computer.[21] Bioprinting is a specific type of biofabrication that involves using a 3D printer to create living tissues by precisely depositing cells and other biomaterials layer by layer. Biofabrication is a broader term that refers to the use of various methods and techniques to create biomaterials, structures, and devices that mimic the properties and functions of natural tissues and organs.[22] 3D bioprinting is the technique that utilises 3D printing technologies in combination with cells, growth factors, and biomaterials to create cellular sheets, scaffolds, tissues, and, hypothetically, whole vascularised organs.[23] Generally, 3D bioprinting utilises the layer-by-layer method to deposit materials

[19] Growth factors, DNA, miRNA, cytokines, exosomes, proteins, etc.

[20] Zhang, Bin, et al. '3D bioprinting: an emerging technology full of opportunities and challenges.' *Bio-Design and Manufacturing* (2018): 1–12.

[21] Zheng, Zhaozhu, et al. '3D Bioprinting of Self-Standing Silk-Based Bioink.' *Advanced healthcare materials* (2018).

[22] Lorenzo Moroni et al., 'Biofabrication: a guide to technology and terminology' (2018) 36(4) *Trends in biotechnology* 384–402.

[23] Murphy, Sean V., and Anthony Atala. '3D bioprinting of tissues and organs.' *Nature biotechnology* 32.8 (2014): 773.

known as 'bioinks'[24] to create tissue-like structures that, it is hoped, will later be commonly used in variety of medical applications, including drug testing, disease modelling, and tissue regeneration.[25] In the future, bioprinting could also be used to create fully functional organs for transplantation, offering a solution to the global shortage of donor organs. Bioinks are a mixture of different biological molecules, and 3D bioprinting uses 'bioinks' in conjunction with human stem cells to bioprint different types of biological products such as tissues and organs. 3D bioprinting is the technology where the two interdisciplinary fields, engineering and biology, come together to develop novel methods and products of medical significance.[26]

While 3D printing has been popular in fields such as engineering, fashion, and even food, recent advances have allowed the 3D printing of biocompatible materials, or bioprinting. Scientists have already printed blood vessels and organ tissues, and even a functional ear.[27] One of the most promising applications of bioprinting is in drug testing and disease modelling. Pharmaceutical companies are using bioprinted tissues to test the safety and efficacy of new drugs in the lab before moving on to clinical trials. This reduces the need for animal testing and speeds up the drug development process. Bioprinted tissues can also be used to model diseases, allowing researchers to study how diseases develop and progress in human tissues. Currently, researchers have used 3D bioprinting to develop cartilage, liver, and cardiac tissues.[28] The University of California, San Diego has announced the development of a new tissue that closely mimics the structure and function of a real human liver tissue.[29] However, the most ambitious goal of bioprinting is to create fully functional human organs. Researchers are already working on bioprinting kidneys, lungs, and hearts, but the complexity of these organs (especially their vascular systems) makes this a daunting challenge. While it may be sometime before bioprinted organs are ready for human transplantation, the progress made so far is remarkable. The potential to

---

[24] Groll, J. et al., 'A Definition of Bioinks and Their Distinction from Biomaterial Inks' (2018) 11 *Biofabrication* 013001.

[25] Kolesky, David B., et al. '3D bioprinting of vascularized, heterogeneous cell-laden tissue constructs.' *Advanced materials* 26.19 (2014): 3124–3130.

[26] Zhang, Yu Shrike, et al. '3D bioprinting for tissue and organ fabrication.' *Annals of biomedical engineering* 45.1 (2017): 148–163.

[27] Che-Ying Kuo et al., 'Repair of Tympanic Membrane Perforations with Customized Bioprinted Ear Grafts Using Chinchilla Models' (2018) 24(5–6) *Tissue Engineering Part A* 527–535; Hooi Yee Ng et al., 'Bioprinting of artificial blood vessels' (2018) 4(2) *International Journal of Bioprinting*.

[28] Nguyen, Duong et al., 'Cartilage Tissue Engineering by the 3D Bioprinting of IPS Cells in a Nanocellulose/Alginate Bioink' (2017) 7 *Scientific Reports* 658; Nguyen, Deborah G. et al., 'Bioprinted 3D Primary Liver Tissues Allow Assessment of Organ-Level Response to Clinical Drug Induced Toxicity In Vitro' (2016) 11 *Plos One;* Jakab, Karoly et al., 'Tissue Engineering by Self-Assembly of Cells Printed into Topologically Defined Structures' (2008) 14 *Tissue Engineering Part A* 413.

[29] Clare Scott, University of California San Diego's 3D Printed Liver Tissue May Be the Closest We've Gotten to a Real Printed Liver, https://3dprint.com/118932/uc-san-diego-3d-printed-liver/.

save lives by providing patients with custom-made organs built from their own cells is driving research forward at a rapid pace.[30]

Advanced research in 3D bioprinting technology and its applications has led to the gradual increase of commercial interests in bioprinting technologies.[31] Organovo, a US-based company specialising in 3D bioprinting, has filed patent applications in different jurisdictions for engineered liver tissue, array thereof, and a method for making it.[32] In this way, different multinational companies, research organisations, and universities have been trying to protect similar kinds of inventions all over the globe.[33] A patent landscape study on 3D bioprinting technology published on a 3D printing industry website reveals a 36% increase in observable patent activity from April 2015 to June 2016.[34]

But as bioprinting advances, so too does the legal uncertainty surrounding it. The creation of living tissues raises fundamental questions about what can be patented and what should remain in the public domain. Unlike traditional medical devices or pharmaceutical products, bioprinted tissues are not entirely synthetic. They involve living cells, which are derived from nature, as well as synthetic materials and engineering processes. This blend of the natural and the man-made complicates the application of patent law, which has traditionally excluded natural phenomena from patent protection. For example, assume that a patent application claims a 3D bioprinted tissue/organ and/or methods/processes of developing the same. The challenge that needs to be addressed is how the traditional patent system is applied and specifically how the subject matter requirement is applied to these claims, as such claims are directed towards a combination of natural (stem cells) and synthetic elements. Similar challenges arise in the context of bioinks.

The challenge of patenting bioprinted tissues is particularly acute when viewed through the lens of different jurisdictions, each with its own approach to patent law. The United States, Europe, and Australia have all developed legal frameworks that address the issue of patentable subject matter in biotechnology, but they do so in different ways. Particularly, each of these regions has developed its own approach to biotechnology patents, shaped by a combination of historical precedents, legal doctrine, and policy considerations. By comparing how each jurisdiction handles the

---

30 Gao, Guifang, Ying Huang, Arndt F. Schilling, Karen Hubbell, and Xiaofeng Cui. 'Organ bioprinting: are we there yet?' (2018). *Advanced healthcare materials* 7, no. 1.

31 Deepak Choudhury, Shivesh Anand and May Win Naing, 'The arrival of commercial bioprinters–Towards 3D bioprinting revolution!' (2018) 4(2) *International Journal of Bioprinting.*

32 *PCT Patent Application No. WO2014151921A1*, filed on 13 March 2014 ('Published' on 25 September 2014).

33 *US Patent Application No. US20160040132A1*, filed on 06 August 2015 ('Published' on 11 February 2016); *US Patent Application No. US20170281828A1*, filed on 24 September 2015 ('Published' on 05 October 2017); *US Patent Application No. US20180171304A1*, filed on 16 June 2016 ('Published' on 21 June 2018).

34 Hornick, John F. and Kai Rajan, *The 3D Bioprinting Patent Landscape Takes Shape as IP Leaders Emerge* (July 7, 2016) 3D Printing Industry https://3dprintingindustry.com/news/3d-bioprinting-patent-landscape-takes-shape-ip-leaders-emerge-84541/.

patentability of bioprinted tissues, this book seeks to identify patterns, inconsistencies, and potential areas for reform. Understanding these differences is crucial for anyone seeking to navigate the complex landscape of bioprinting patents, particularly as these technologies are developed and utilised by international research teams and routinely cross jurisdictional boundaries, underscoring the need for consistency to avoid confusion and support innovation.

In the United States, the *Mayo* and *Myriad* decisions have set a high bar for patenting biological inventions. In *Mayo Collaborative Services v. Prometheus Laboratories*, the U.S. Supreme Court ruled that a diagnostic method for measuring metabolite levels in the human body was not patentable because it relied on natural biological processes.[35] Similarly, in *Myriad*, the Court ruled that isolated DNA sequences could not be patented because they were naturally occurring.[36] These decisions have narrowed the scope of patentable subject matter in biotechnology, making it more difficult for companies to obtain patents on biological inventions that involve natural materials. However, the courts have left some room for inventions that involve significant human intervention, and bioprinting may fall into this category if the synthetic aspects of the technology are emphasised in patent claims.

In Europe, the European Patent Convention (EPC) provides a framework for determining patent eligibility, but it also includes explicit exclusions for inventions that violate *ordre public* or morality. These exclusions have been invoked in cases involving human stem cells, raising concerns about the patentability of bioprinted tissues that rely on stem cell technologies. The EPC also requires that inventions involve an inventive step and be susceptible to industrial application, which may be difficult to demonstrate for bioprinted tissues that are still in the experimental stage.

In contrast to both of the previous jurisdictions, Australia's approach to biotechnology patents is shaped by the 'manner of manufacture' test, which requires that an invention creates an 'artificially created state of affairs' with economic utility. This test, which dates back to the High Court's 1959 decision in *National Research Development Corporation v. Commissioner of Patents (NRDC)*, has been applied in a range of contexts, including biotechnology.[37] The test provides more flexibility for patenting biotechnological inventions than the U.S. approach, and it could potentially allow for the patenting of bioprinted tissues, provided they are sufficiently modified from their natural state.

The differences between these jurisdictions highlight the complexity of patenting bioprinting technologies. A bioprinted tissue that qualifies for patent protection in one country may be ineligible in another, depending on how the courts interpret the patentable subject matter requirements. This creates uncertainty for inventors and companies working in the field of bioprinting, who must navigate a patchwork of legal rules and precedents as they seek to protect their innovations.

Given the complexities of patenting bioprinting innovation, this book explores whether patentable subject matter requirement poses a hurdle to patenting bioprinting

[35] *Mayo Collaborative Services v Prometheus Laboratories, Inc*, 566 US 66 (2012).

[36] *Association for Molecular Pathology v Myriad Genetics Inc*, 133 US 2107 (2013).

[37] *National Research Development Corporation v Commissioner of Patents* (1959) 102 CLR 252.

inventions, specifically bioinks and bioprinted tissues. This central patentability issue revolves around patentable subject matter, a fundamental requirement in patent law that determines what types of inventions are eligible for patent protection. Traditionally, patent law excludes natural phenomena and products of nature from protection, meaning that biological materials in their natural state cannot be patented. However, the blending of natural and synthetic materials in bioprinting complicates this analysis. To what extent can human intervention transform a natural material into something patentable? And how do courts in different jurisdictions interpret and apply this principle? How are patent offices currently assessing and responding to patent claims related to 3D bioprinted tissues and bioinks?

This book explores the patentability of bioprinted tissues and bioinks through a combination of doctrinal analysis and empirical research, examining how patent laws in the United States, Europe, and Australia are being applied to bioprinting technologies. Specifically, this book examines the capacity of the traditional patent system to adapt to the challenges of assessing patentability of 3D bioprinting innovation, by assessing whether specific subject matter related to 3D bioprinted tissues and bioinks fall within the scope of patentable subject matter and thus considered to be patentable. In doing so, it explores the current application of the patentability requirements to bioprinted tissues and bioinks, with a detailed examination of the patentable subject matter requirement.

Each chapter of the book builds on the central inquiry and aims to address one or more of the above questions that stemmed out of the central research question. Chapter 2 provides a detailed background on 3D bioprinting technologies, explaining how the technology works and the legal challenges it introduces. Chapters 3, 4, and 5 offer a comparative analysis of the patent laws in Australia, the United States, and Europe, respectively, focusing on how each jurisdiction applies its subject matter requirements to bioprinted inventions. Chapter 6 presents an empirical analysis of patent prosecution data, revealing how patent offices are responding to bioprinting patent applications, particularly in terms of objections and amendments based on subject matter eligibility. Finally, Chap. 7 synthesises these findings and proposes a framework for how bioprinting patents might be more effectively managed, offering practical recommendations for overcoming the subject matter hurdles that exist.

The analysis presented in this book is not just about whether bioprinting technologies fit within the current boundaries of patent law. It is about how the law itself must evolve to accommodate these new technologies. As bioprinting continues to develop, the legal system will need to adapt, not just to protect these innovations, but also to ensure that they serve the greater good. The future of intellectual property in the age of bioprinting hinges on how patent law navigates the complex relationship between fostering innovation and ensuring widespread access to potentially life-saving technologies. In this book, I aim to chart a course through these complex matters, offering insights into how patent systems might evolve to address the unique challenges posed by bioprinting while continuing to serve their fundamental purpose of promoting scientific and technological progress.

Chapter 2
3D Bioprinting Innovation: Past, Present, and Future

2.1 Introduction

This chapter sets the scene for this book by elaborating on the technological aspects of bioprinting, including the historical context of bioprinting, how the technology has evolved and increased in complexity, and why there is an ongoing need for innovation in this field to understand the demand for patents. In addition, this chapter also provides an overview of advancements in the field of bioprinting with a specific emphasis on how bioprinting technology is in a constant state of innovation leading to inventions with complex subject matter. The initial sections of this chapter provide a historical context of bioprinting technology and how the technology has evolved. This chapter then discusses the complex composition of a bioink (a basic structural element of a 3D bioprinted tissue or scaffold) and sheds light on different steps involved in 3D bioprinting. The remaining sections of this chapter discuss the recent advancements in bioprinting technology with a focus on different types of bioprinting techniques (such as inkjet bioprinting, stereolithographic bioprinting, extrusion bioprinting, and laser-assisted bioprinting) used for printing bioink into the form of a bioprinted tissue or scaffold. In doing so, this chapter explores the technical aspects of bioprinting and stands as a foundation to some of the arguments of this book on complexity of bioprinting subject matter and how recent advances in bioprinting technology may challenge the traditional patent system.

P. Devarapalli, *Bioinked Boundaries*, https://doi.org/10.1007/978-981-96-2591-8_2

2.2 Historical Context and Evolution of Bioprinting

3D bioprinting utilises a layer-by-layer method to deposit materials such as bioinks to create tissue-like structures that are later used in medical applications.[1] Bioprinting is an evolved branch of tissue engineering and regenerative medicine. The aim of traditional tissue engineering and regenerative medicine is to assemble functional constructs that restore, maintain, or improve damaged tissues. Bioprinting technology provide new techniques for engineering tissue, by three dimensionally building a tissue or organ. This offers potential advantages in that conventional tissue engineering techniques do not permit the building or printing of three-dimensional structures.[2] Traditionally, stem cells are cultured in two-dimensional (2D) plastic culture plates as a monolayer. However, 2D culture methods are essentially ineffective, costly, and labour-intensive, and only achieve limited stem cell expansion. Additionally, the complexity required to mimic the 3D environment of a stem cell niche that controls stem cell fate is lacking in 2D culture conditions.[3] Bioprinting has evolved to address problems that traditional tissue engineering and regenerative medicine technologies are not able to address. Specifically, the need for support structures for structural proliferation of stem cells has driven bioengineers to develop a technology that can address these issues.[4] Additionally, 3D bioprinted tissues are addressing a need for authentic three-dimensional structures that mimic body tissues for the purposes of drug delivery and cancer studies, aimed at understanding how the cells respond to potential drugs in the same way that they do in a patient.[5]

In 1984, Charles developed and patented a stereolithography method that allowed him to create models from resins layer by layer.[6] Four years later, in 1988, Klebe developed a bioprinting method for micropositioning cells, and for the construction of 2D and 3D synthetic tissues.[7] Biomaterial was used to regenerate tissue for the

---

[1] D. B. Kolesky et al., '3D Bioprinting of Vascularized, Heterogeneous Cell-Laden Tissue Constructs' (2014) 26(19) *Advanced Materials* 3124–3130.

[2] François Berthiaume, Timothy J. Maguire and Martin L. Yarmush, 'Tissue engineering and regenerative medicine: history, progress, and challenges' (2011) 2 *Annual review of chemical and biomolecular engineering* 403–430.

[3] "Benefits Of Bioprinting—Brinter™ Bioprinter—3D Bioprinting", *Brinter—3D Bioprinting Solutions* (Webpage, 2022) https://www.brinter.com/benefits-of-bioprinting/.

[4] Christian Mandrycky et al., '3D bioprinting for engineering complex tissues' (2016) 34(4) *Biotechnology advances* 422–434.

[5] Aishwarya Satpathy et al., 'Developments with 3D bioprinting for novel drug discovery' (2018) 13(12) *Expert opinion on drug discovery* 1115–1129; Arpita P. Tiwari et al., 'Bioink: a 3D-bioprinting tool for anticancer drug discovery and cancer management' (2021) 26(7) *Drug Discovery Today* 1574–1590.

[6] *Apparatus for production of three-dimensional objects by stereolithography*, Priv L No US, ch 3D Systems Inc, US4575330B1 (1984) ('*Apparatus for production of three-dimensional objects by stereolithography*').

[7] Robert J. Klebe, 'Cytoscribing: a method for micropositioning cells and the construction of two- and three-dimensional synthetic tissues' (1988) 179(2) *Experimental cell research* 362–373.

first time ever in 1996, and several years later in 1999, Odde and Renn used laser-assisted bioprinting to deposit living cells for constructing analogues with complex anatomical structures.[8] In 2001, a bladder-shaped scaffold was bioprinted, and donor cells seeded into it.[9] Through 2002 and 2003, Thomas and his team developed an inkjet bioprinting technique using cells that had a high viability and patented this technology.[10] A year later, the same team developed a cell-loaded bioprinting technique for rapid prototyping of tissue-engineering constructs using stereolithography.[11] In 2006, an electrodynamic jetting technique enabled depositing of living cells.[12]

These events led to the development of the first commercial bioprinter, NovoGen MMX, in 2009.[13] This device enabled bioengineers to construct the first biodegradable blood vessel without the use of cell scaffolds.[14] Many new bioprinting products were introduced in the years that followed, including scaffold-free vascular constructs,[15] articular cartilage,[16] artificial liver,[17] heart valves,[18] and cornea.[19] A bioprinted cardiac model was developed by the Feinberg lab at Carnegie Mellon University in 2020 that replicates the natural feel, elasticity, and mechanical characteristics of cardiac tissue.[20] In 2021, miniature human kidneys were successfully bioprinted in the lab by scientists at the Australian Murdoch Children's Research

[8] Yoshihiro Shigemasa and Saburo Minami, 'Applications of chitin and chitosan for biomaterials' (1996) 13(1) *Biotechnology and Genetic Engineering Reviews* 383–420; David J. Odde and Michael J. Renn, 'Laser-guided direct writing for applications in biotechnology' (1999) 17(10) *Trends in biotechnology* 385–389.

[9] Anthony Atala, 'Tissue engineering in urology' (2001) 2(1) *Current urology reports* 83–92.

[10] *Ink-jet printing of viable cells*, Priv L No US, ch Clemson University Research Foundation, (2006) ('*Ink-jet printing of viable cells*').

[11] Busaina Dhariwala, Elaine Hunt and Thomas Boland, 'Rapid prototyping of tissue-engineering constructs, using photopolymerizable hydrogels and stereolithography' (2004) 10(9–10) *Tissue engineering* 1316–1322.

[12] Suwan N. Jayasinghe, Amer N. Qureshi and Peter A. M. Eagles, 'Electrohydrodynamic jet processing: an advanced electric-field-driven jetting phenomenon for processing living cells' (2006) 2(2) *Small* 216–219.

[13] *Organ printing: a novel tissue engineering paradigm*) (5th European Conference of the International Federation for Medical and Biological Engineering) ('*Organ printing: a novel tissue engineering paradigm*').

[14] Ibid.

[15] Cyrille Norotte et al., 'Scaffold-free vascular tissue engineering using bioprinting' (2009) 30(30) *Biomaterials* 5910–5917.

[16] Xiaofeng Cui et al., 'Direct human cartilage repair using three-dimensional bioprinting technology' (2012) 18(11–12) *Tissue Engineering Part A* 1304–1312.

[17] Igor M. Sauer, Nathanael Raschzok and Peter Neuhaus, 'The artificial liver, in vivo tissue engineering, and organ printing' (2014) *Textbook of organ transplantation* 545–553.

[18] Bin Duan et al., 'Three-dimensional printed trileaflet valve conduits using biological hydrogels and human valve interstitial cells' (2014) 10(5) *Acta biomaterialia* 1836–1846.

[19] Abigail Isaacson, Stephen Swioklo and Che J. Connon, '3D bioprinting of a corneal stroma equivalent' (2018) 173 *Experimental eye research* 188–193.

[20] Eman Mirdamadi et al., 'FRESH 3D bioprinting a full-size model of the human heart' (2020) 6(11) *ACS Biomaterials Science & Engineering* 6453–6459.

Institute in association with Organovo Inc.[21] Finally in 2022, a patient was the recipient of a 3D bioprinted ear implant made from the patient's own living cells.[22] Recent advances in bioprinting have helped the technology expand its boundaries to tissue and organ development.[23] Prediction of drug efficacy and toxicity plays a crucial role in the development of efficient drugs. Due to their ability to mimic the biological characteristics of native tissues, 3D bioprinted tissue models can be used in in vitro preclinical drug discovery processes for predicting drug efficacy and toxicity.[24]

Researchers at the Massachusetts Institute of Technology (MIT) have, for the first time, reported the capacity to undertake 4D printing. The fourth dimension that they have described is 'transformation of printed structures over time.'[25] 3D bioprinting provides a structural framework with specific growth environment to the living cells to proliferate and grow into a specific tissue or organ. 4D bioprinting uses advanced bioprinting strategies and creates structures that provide time-dependent shape changes in the bioprinted constructs.[26] Biofabricating 'external stimuli'-based responsive/transformative 3D constructs is considered as the additional fourth dimension in 4D bioprinting.[27] Using these techniques, different biomaterials have been developed that can transform or respond to temperature, magnetic field, and water.[28] The changes occurring in the 4D bioprinted structures are in respect of shape, size, and pattern.[29] In this regard, some recent research related to bioprinting is directed towards developing bioinks and bioprinting techniques that are useful for 4D bioprinting.[30] In this manner, bioprinting technology has evolved into different forms and is in a constant state of innovative flux to address emerging issues in tissue engineering and medicine.

---

[21] Christine Horejs, 'Bioprinted mini kidneys' (2021) 6(1) *Nature Reviews Materials* 6 6; Kynan T. Lawlor et al., 'Cellular extrusion bioprinting improves kidney organoid reproducibility and conformation' (2021) 20(2) *Nature materials* 260–271.

[22] Sakharkar, Ashwini, "3Dbio Implants 3D-Bioprinted Ear In A First-In-Human Clinical Trial", *Inceptive Mind* (Webpage, 2022) https://www.inceptivemind.com/3dbio-implants-aurinovo-3d-bioprinted-ear-first-human-clinical-trial/25005/.

[23] S. V. Murphy and A. Atala, '3D Bioprinting of Tissues and Organs' (2014) 32(8) (2014/08/06) *Nature Biotechnology* 773–85.

[24] Weijie Peng et al., '3D Bioprinting for Drug Discovery and Development in Pharmaceutics' (2017) 57 *Acta biomaterialia* 26–46.

[25] S. Tibbits, '4D Printing: Multi-Material Shape Change' (2014) 84(1) *Architectural Design* 116–121.

[26] Y. C. Li et al., '4D Bioprinting: The Next-Generation Technology for Biofabrication Enabled by Stimuli-Responsive Materials' (2017) 9(1) *Biofabrication.*

[27] N. Ashammakhi et al., 'Advances and Future Perspectives in 4D Bioprinting' (2018) 13(12) *Biotechnology Journal.*

[28] B. Gao et al., '4D Bioprinting for Biomedical Applications' (2016) 34(9) *Trends in Biotechnology* 746–756.

[29] J. An, C. K. Chua and V. Mironov, 'A perspective on 4D bioprinting' (2016) 2(1) *International Journal of Bioprinting* 3–5.

[30] A. Kirillova et al., '4D Biofabrication Using Shape-Morphing Hydrogels' (2017) 29(46) *Advanced Materials.*

2.3 Bioinks and Bioprinted Tissues or Scaffolds

Bioprinted tissues or scaffolds are composed of the natural elements (such as living cells, bioactive molecules) in natural tissue as well as artificial elements to support or promote the growth of natural elements (such as growth factors, biomaterial, or biopolymer gel) in an artificial environment.[31] For example, a group of researchers have bioprinted a meniscus, a piece of cartilage in the knee that prevents the shinbone and thighbone from grinding against each other.[32] Meniscus is made up of fibrochondrocytes (cartilage-forming cells) with few myofibroblasts and endothelial cells.[33] These cells can come from donors whose cell lines have been duplicated in a lab, or they can come from a patient's own tissue to generate a customised meniscus that will be less likely to be rejected by their body. After the bioprinting process is complete, the cells in the bioprinted tissue will begin to function similarly to cells in natural tissue, by communicating, exchanging nutrients, and proliferating.[34]

Vascularisation is the process of growing blood vessels into a tissue to improve oxygen and nutrient supply. Bioprinting vascularised tissues is one of the most intractable challenges of bioprinting technology (and tissue engineering more generally) enabling the production of sizeable tissue and organ substitutes that contain living cells.[35] A recently published study considered the potential and processability of spheroids (3D cell aggregates that can mimic tissue) with a capillary-like network for usage as micron-sized pre-vascularised units in bioprinted constructs. In this study, scientists combined endothelial cells (cells found in the inside lining of blood vessels, lymph vessels, and the heart) with fibroblasts (cells that contribute to the formation of a fibrous cellular material that supports and connects other tissues or organs in the body) and adipose tissue-derived mesenchymal stem cells (cells with the capacity for self-renewal) as supporting cells to create pre-vascularised spheroids.[36] Similarly, a group of researchers have created aligned cardiac tissues by bioprinting organ building blocks consisting of human induced pluripotent stem cell-derived cardiomyocytes (cells responsible for generating contractile force in the

[31] Hannah A. Strobel, Sarah M. Moss and James B. Hoying, 'Mechanical considerations of bioprinted tissue' (2020) 6 *Frontiers in Mechanical Engineering* 568726.

[32] Giuseppe Filardo et al., 'Patient-specific meniscus prototype based on 3D bioprinting of human cell-laden scaffold' (2019) 8(2) *Bone & joint research* 101–106.

[33] The peripheral two-thirds of the meniscus solely consist of type I collagen, whereas type II collagen comprises a significant portion of the fibrillar collagen in the inner meniscus; Peter C. M. Verdonk et al., 'Characterisation of human knee meniscus cell phenotype' (2005) 13(7) *Osteoarthritis and Cartilage* 548–560.

[34] Filardo et al. (n 32).

[35] François A. Auger, Laure Gibot and Dan Lacroix, 'The pivotal role of vascularization in tissue engineering' (2013) 15(1) *Annual Review of Biomedical Engineering* 177–200.

[36] Lise De Moor et al., 'Engineering microvasculature by 3D bioprinting of prevascularized spheroids in photo-crosslinkable gelatin' (2021) 13(4) *Biofabrication* 045021.

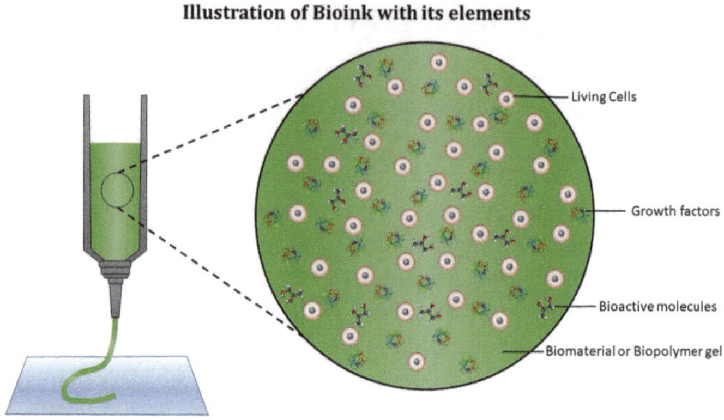

Fig. 2.1 Graphic illustration of bioink visualising living cells, growth factors, and bioactive molecules suspended in biomaterial or biopolymer gel

intact heart).[37] This study opens up new possibilities for producing functioning heart tissue with a high cell density and complex cellular alignment.

Bioinks are the materials that are used to develop living and functional tissues with the help of 3D bioprinting technology. As visualised in Fig. 2.1, the basic composition of bioink includes (but is not limited to) living cells, bioactive molecules, growth factors, and biomaterial or biopolymer gels that act as carriers and envelop the living cells, providing a structural configuration to the bioprinted cells and enabling them to reproduce quickly.[38] In addition, the biomaterial also provides a protective environment to the living cells during the bioprinting process so that bioprinted cells can stay active to further multiply. The bioink's physical, chemical, and biological properties are essential for printability, spatial stability, shape accuracy, and cell sustainability before, during, and after the bioprinting process.[39]

An ideal bioink should possess good mechanical and biological properties, which are essential to ensure proper functionality of the bioprinted tissues developed using specific bioink.[40] Materials or polymer gels used in bioinks that give structural configuration to the bioinks can be of natural or synthetic origin.[41] Specifically, biomaterials or biopolymer gels are mostly derived from plants or animals. The major reasons behind using plants or animals as a source of biomaterials for bioinks are that they

[37] John Ahrens et al., 'Programming Cellular Alignment in Engineered Cardiac Tissue via Bioprinting Anisotropic Organ Building Blocks' (2022) *Advanced Materials* 2200217.

[38] M. Hospodiuk et al., 'The Bioink: A Comprehensive Review on Bioprintable Materials' (2017) 35(2) *Biotechnology Advances* 217–239.

[39] K. Holzl et al., 'Bioink Properties Before, During and After 3D bioprinting' (2016) 8(3) *Biofabrication.*

[40] P. Selcan Gungor-Ozkerim et al., 'Bioinks for 3D bioprinting: an overview' (2018) 6(5) *Biomaterials science* 915–946.

[41] D. Chimene et al., 'Advanced Bioinks for 3D Printing: A Materials Science Perspective' (2016) 44(6) *Annals of Biomedical Engineering* 2090–2102.

possess innate biocompatibility (the ability of a material to perform with an appropriate host response in a specific application) and can exhibit good efficiency in cell communication.[42] Bioinks developed using collagen-based biomaterials derived from natural sources are useful for printing collagen-based tissues structures such as cartilage or skin.[43] Similarly, fibrin-based biomaterials are used in bioinks that are exclusively used for printing cardiovascular tissues.[44] Alternatively, synthetically produced materials used in bioinks offer better mechanical properties and bio-printability. Depending on the requirement, natural or synthetic biomaterials, either separately or in combination, can be used. Bioinks comprised of synthetic polymers are also used to construct different tissue types even though their chemical and structural properties are different in comparison with natural biomaterials.[45] However, these synthetically based bioinks come with some limitations such as poor cell encapsulation and cell adhesion.[46] In this regard, biomaterials derived from nature are preferred for the development of bioinks.[47]

Bioinks have increasingly gained importance and play a crucial role in bioprinting processes. Recent advancements in bioprinting have given rise to development of novel bioinks that possess better biocompatibility, biodegradability, and printability.[48] Biocompatibility is the capacity of a bioink to maintain high levels of cell viability to promote cell growth and proliferation and to preserve the population of cells.[49] Biodegradability enables bioinks to integrate readily into natural tissues. Particularly, biodegradable bioinks serve as temporary templates, engage in cell interaction, merge with natural tissues, and support tissue repair.[50] Printability refers to the capacity of the bioprinting process to create and sustain reproducible 3D scaffolds from bioink.[51]

---

[42] D. Zhou et al., 'Bioinks for Jet-based Bioprinting' (2019) *Bioprinting* e00060.

[43] Egor O. Osidak et al., 'Viscoll Collagen Solution as a Novel Bioink for Direct 3D Bioprinting' (2019) 30(3) *Journal of Materials Science-Materials in Medicine.*

[44] S. A. Kumar et al., 'A Visible Light Crosslinkable, Fibrin-Gelatin Based Bioprinted Cardiac Construct with Human Cardiomyocytes and Fibroblasts' (2019) *ACS Biomaterials Science and Engineering.*

[45] A. Skardal, J. Zhang and G. D. Prestwich, 'Bioprinting Vessel-Like Constructs using Hyaluronan Hydrogels Crosslinked with Tetrahedral Polyethylene Glycol Tetracrylates' (2010) 31(24) *Biomaterials* 6173–6181.

[46] P. Camacho et al., 'Materials as Bioinks and Bioink Design', *3D Bioprinting in Medicine* (Springer, 2019) 67–100.

[47] J. Gopinathan and I. Noh, 'Recent Trends in Bioinks for 3D Printing' (2018) 22 *Biomaterials Research* 11–11.

[48] Shen Ji and Murat Guvendiren, 'Recent Advances in Bioink Design for 3D Bioprinting of Tissues and Organs' (2017) 5 *Frontiers in Bioengineering and Biotechnology* 23–23.

[49] Matthew L. Bedell et al., 'A high-throughput approach to compare the biocompatibility of candidate bioink formulations' (2020) 17 *Bioprinting* e00068.

[50] Yiting Ze et al., 'Biodegradable Inks in Indirect Three-Dimensional Bioprinting for Tissue Vascularization' (2022) *Frontiers in Bioengineering and Biotechnology* 461.

[51] Saman Naghieh and Xiongbiao Chen, 'Printability–A key issue in extrusion-based bioprinting' (2021) 11(5) *Journal of Pharmaceutical Analysis* 564–579.

The most commonly used bioinks for tissue bioprinting are cell-laden hydrogels (scaffolds that encapsulate cells within their crosslinked network to mimic native tissue-like structure and function),[52] decellularised extracellular matrix (ECM) (a biomaterial that preserves a tissue's native environment) based gels,[53] and cell suspensions (a type of bioink in which single cells multiply as they lie suspended in a predefined cell media).[54] Cell-laden hydrogels are specifically used in most bioprinting techniques due to their programmable characteristics and their capacity to recapitulate the cell microenvironment.[55] Bioinks based on decellularised extracellular matrix are actively evolving due to their intrinsic bioactivity and flexibility to become integrated into the complex bioink formulation.[56] In the same way, cell suspension bioinks based on cell aggregates are a feasible alternative for creating scaffold-free 3D bioprinted tissue constructs.[57]

2.4 Bioprinting—What Are the Steps Involved?

There are several critical steps in the process of bioprinting a tissue or scaffold. The first stage in creating a tissue scaffold that accurately represents the composition and structure of the tissue of interest is structural design. In this stage, a detailed understanding of the targeted tissue structure is required to appropriately bioprint a structure that can mimic the natural tissue. As visualised in Fig. 2.2, the next steps are creating a blueprint of the desired tissue or organ using Magnetic Resonance Imaging (MRI)[58] or Computed Tomography (CT) scans from a patient.[59] Once the blueprint data is obtained, the next step is converting the tissue or organ blueprint

[52] Sarah M. Hull, Lucia G. Brunel and Sarah C. Heilshorn, '3D bioprinting of cell-laden hydrogels for improved biological functionality' (2022) 34(2) *Advanced Materials* 2103691.

[53] D. Choudhury et al., 'Organ-Derived Decellularized Extracellular Matrix: A Game Changer for Bioink Manufacturing?' (2018) 36(8) *Trends in Biotechnology* 787–805.

[54] Y. Hussain and J. T. Butcher, 'Bioprinting Cardiovascular Organs', *3D Printing Applications in Cardiovascular Medicine* (Elsevier, 2018) 163–187.

[55] S. Ji, E. Almeida and M. Guvendiren, '3D Bioprinting of Complex Channels within Cell-Laden Hydrogels' (2019) 95 *Acta Biomaterialia* 214–224.

[56] Byoung Soo Kim et al., 'Decellularized Extracellular Matrix: A Step Towards the Next Generation Source for Bioink Manufacturing' (2017) 9(3) *Biofabrication.*

[57] Ibrahim T. Ozbolat, 'Scaffold-Based or Scaffold-Free Bioprinting: Competing or Complementing Approaches?' (2015) 6(2) *Journal of Nanotechnology in Engineering and Medicine* 024701.

[58] Magnetic resonance imaging is a medical imaging technology that creates detailed images of your organs and tissues using a magnetic field and computer-generated radio waves.

[59] Computed tomography also known as computerised axial tomography or computerised tomography. It is a type of imaging that employs specialised X-ray equipment to make detailed images, or scans, of internal organs.

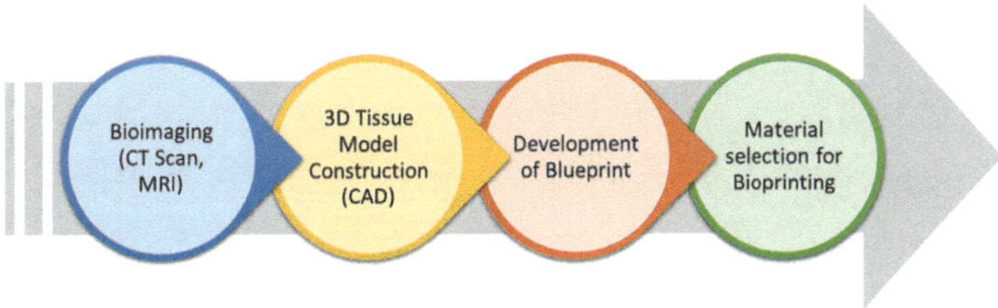

Fig. 2.2 A stepwise illustration of pre-processing steps involved in bioprinting

data into bioprinter readable format which primarily utilises Computer Aided Design (CAD) or Computer Aided Manufacturing (CAM)[60] format.

After converting the blueprint data into bioprinter readable format, a process of identifying and isolating specific living cells and differentiating the cells into organ or tissue-specific cells is undertaken: this involves synthesis of biomaterials that can support the tissue's desired cellular and extracellular functions.

Once the required cells are isolated, the next step includes the preparation of bioink with the combination of differentiated cells, blood vessel cells, growth factors, and supporting biofabrication material where the biomaterial chosen must be compatible not only with the functions of the selected cells and tissues, but also with the bioprinting process. Once the bioink is prepared, the next step is proliferation[61] of stem cells in the bioink, and after the proliferation stage the bioink is loaded into the printheads and the bioprinting process is initiated using the software connected to the bioprinter. Once the bioprinting process is completed, the bioprinted tissue that was developed is placed in a bioreactor. During tissue maturation, bioreactors are used to provide an in vitro physiological environment that is specific to the tissue.[62] It can be used for applications such as disease study, drug discovery, and transplantation as shown in Fig. 2.3.[63]

As noted, currently bioprinting can be used to print tissues and organ models for medical and pharmaceutical research.[64] Specifically, the present applications of bioprinting include bioprinted tissue models for cancer research, drug screening,

---

[60] Computer-aided design or Computer-aided manufacturing is the use of computers to assist in the creation, modification, analysis, or optimisation of a product design. This program is used to improve the designer's efficiency, the quality of the design, communication through documentation, and the creation of a database for manufacturing different types of products including bioprinted products.

[61] Stem cell proliferation describes an increase in cell number caused by cell growth and division into two identical daughter cells.

[62] Jordan Rosser and Daniel J. Thomas, 'Bioreactor processes for maturation of 3D bioprinted tissue', *3D Bioprinting for reconstructive surgery* (Elsevier, 2018) 191–215.

[63] Jipeng Li et al., 'Recent Advances in Bioprinting Techniques: Approaches, Applications and Future Prospects' (2016) 14(1) *Journal of translational medicine* 271.

[64] W. Jamroz et al., '3D Printing in Pharmaceutical and Medical Applications—Recent Achievements and Challenges' (2018) 35(9) *Pharmaceutical Research.*

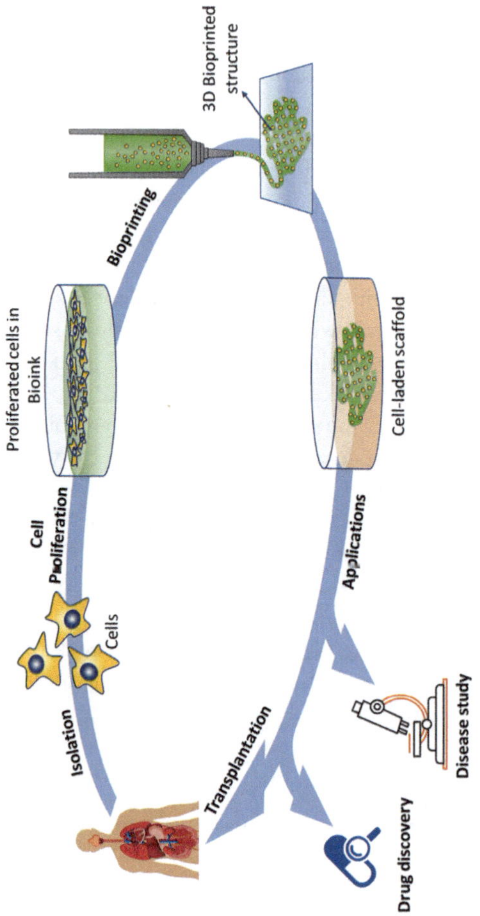

Fig. 2.3 Graphical illustration of different steps involved in bioprinting and its applications

and toxicology screening.[65] For example, a liver-like model utilising living cells and special hydrogel has been reported to provide a means of high-throughput screening of drugs to assess liver toxicity in vitro.[66] The Wyss Institute's multidisciplinary bioprinting research has led to development of a multi-material 3D bioprinting method that produces vascularised kidney tissues made up of living human cells that are nearly ten times thicker than previously engineered tissues and can maintain their architecture and function for up to 6 weeks. The approach employs a customised, printed silicone mould to house and align the printed tissue on a chip. A grid of larger vascular channels containing living endothelial cells in silicone ink is printed inside the mould, which is then layered with a self-supporting bioink containing living mesenchymal stem cells.[67]

2.5 Bioprinting Techniques and Recent Advances

2.5.1 Extrusion-Based Bioprinting

Extrusion-based bioprinting, also known as solid freeform bioprinting, is a set of processes for creating tissues or scaffolds by sequentially delivering bioink in a layer-by-layer fashion based on CAD data.[68] Extrusion-based bioprinting is considered to be one of the most preferred bioprinting methods to bioprint a tissue or scaffold. Bioprinting high viscous bioinks requires high pressure, and this method not only enhances printing resolution by regulating bioprinting pressure, but it is also compatible with a variety of different materials.[69]

As visualised in Fig. 2.4, extrusion-based bioprinting is performed in different ways. There are two main types of extrusion-based bioprinting techniques: pneumatic and mechanical extrusion. Figure 2.4a demonstrates pneumatic extrusion, which extrudes bioink from a nozzle using a compressed gas such as air or nitrogen, with pressure controlling both the flow rate and the volume of the extruded bioink. In contrast, mechanical extrusion extrudes bioink using mechanical force delivered by applying a direct force using a piston (Fig. 2.4b) or by a rotational screw

---

65 "Benefits Of Bioprinting—Brinter™ Bioprinter—3D Bioprinting", *Brinter—3D Bioprinting Solutions* (Webpage, 2022) https://www.brinter.com/benefits-of-bioprinting/.

66 S. C. Ramaiahgari et al., 'A 3D In vitro Model of Differentiated HepG2 Cell Spheroids with Improved Liver-like Properties for Repeated Dose High-Throughput Toxicity Studies' (2014) 88(5) *Archives of Toxicology* 1083–1095; ibid.

67 David B. Kolesky et al., 'In vitro human tissues via multi-material 3-D bioprinting' (2018) 46(4) *Alternatives to Laboratory Animals* 209–215.

68 Tao Jiang et al., 'Extrusion bioprinting of soft materials: An emerging technique for biological model fabrication' (2019) 6(1) *Applied Physics Reviews* 011310.

69 Liqun Ning and Xiongbiao Chen, 'A brief review of extrusion-based tissue scaffold bio-printing' (2017) 12(8) *Biotechnology journal* 1600671.

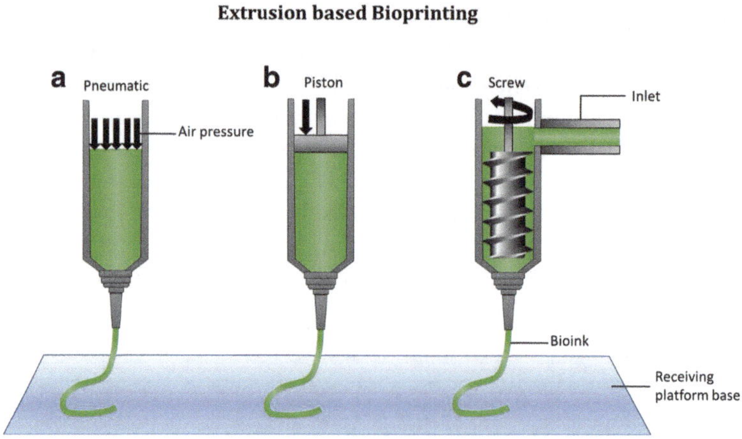

Fig. 2.4 Schematic visualisation of extrusion-based bioprinting. **a** Pneumatic extrusion, **b** Mechanical extrusion by piston, and **c** Mechanical extrusion by rotational screw

(Fig. 2.4c). These methods have different capacities; for example, pneumatic extrusion bioprinters have simpler printheads (a printhead is a component in a bioprinter that houses bioink cartridges) than mechanical extrusion bioprinters, despite being inferior in precision control.[70]

In comparison with the other bioprinting methods, extrusion-based bioprinting supports a variety of fluid viscosities of bioinks. Viscosity of a fluid is a measure of its resistance to deformation at a given rate. In terms of providing structural support for bioprinted tissue scaffolds, higher viscosity materials are preferable, while lower viscosity materials work well for cellular bioactivities such as having an effect on or eliciting a response from living tissue.[71] As a result, in extrusion-based methods significant consideration must be given to the synthesis of materials in order to maintain both mechanical stability of tissue scaffold constructions and cell functions.[72] Specifically in extrusion-based bioprinting, controlling the bioprinting process or the nozzle diameter can improve the mechanical stability of structures by increasing the diameter of dispensed bioink. On the other hand, thick filaments, and lower porosities, reduce cellular diffusion (movement of biomolecules into or

[70] Yu Shrike Zhang et al., '3D extrusion bioprinting' (2021) 1(1) *Nature Reviews Methods Primers* 1–20.

[71] Liming Bian et al., 'The influence of hyaluronic acid hydrogel crosslinking density and macromolecular diffusivity on human MSC chondrogenesis and hypertrophy' (2013) 34(2) *Biomaterials* 413–421; Christopher James Little, Nahshon Kenneth Bawolin and Xiongbiao Chen, 'Mechanical properties of natural cartilage and tissue-engineered constructs' (2011) 17(4) *Tissue Engineering Part B: Reviews* 213–227.

[72] Xiaohong Wang, Boaz Lloyd Rijff and Gilson Khang, 'A building-block approach to 3D printing a multichannel, organ-regenerative scaffold' (2017) 11(5) *Journal of Tissue Engineering and Regenerative Medicine* 1403–1411.

outside a cell) and thus prevent the exchange of nutrients and metabolic excrements.[73] Diffusion barriers can be eliminated by reducing the diameter of extruded bioink, but this would weaken its mechanical properties. Adjusting the mechanical properties of scaffolds by regulating material varieties, material viscosities, or the bioprinting process is essential for structural stability and cellular functionality (in the bioprinted tissue) as different cell types require varied growth conditions.[74]

2.5.2 *Inkjet-Based Bioprinting*

Inkjet-based bioprinting works by depositing bioink in the form of droplets to form a desired three-dimensional structure.[75] This is comparatively different from the output provided by extrusion-based bioprinting as extrusion-based bioprinting prints bioink in the form of filaments rather than droplets.[76] Inkjet-based bioprinting is capable of printing living cells, at micro and nanometer resolution, similar to the construction of natural biological tissues.[77] Inkjet bioprinting has been commonly classified into thermal, piezoelectric, and electromagnetic techniques as visualised in Fig. 2.5.[78] In thermal inkjet printers, heating elements heat the printhead, forming small air bubbles that generate pressure pulses that release ink droplets from the nozzle.[79] In the piezoelectric inkjet technique, a voltage pulse is applied to a piezoelectric material, causing the shape to deform, which causes the ink volume in the pressure chamber to change, generating a pressure wave that perpetuates towards the nozzle, breaking the liquid into droplets at regular intervals.[80] Similarly, electromagnetic inject bioprinting uses electromagnetic forces to expel successive drops of bioink onto a substrate.[81] One of the advantages of inkjet bioprinting is the use of multiple

[73] Jinshu Mao et al., 'Study of novel chitosan-gelatin artificial skin in vitro' (2003) 64(2) *Journal of Biomedical Materials Research Part A: An Official Journal of The Society for Biomaterials, The Japanese Society for Biomaterials, and The Australian Society for Biomaterials and the Korean Society for Biomaterials* 301–308.

[74] Ning and Chen (n 69).

[75] Kyle W. Binder et al., 'Drop-on-Demand Inkjet Bioprinting: A Primer' (2011) 6(01) *Gene Therapy and Regulation* 33–49.

[76] See Figs. 2.4 and 2.5.

[77] Makoto Nakamura et al., 'Biocompatible Inkjet Printing Technique for Designed Seeding of Individual Living Cells' (2005) 11(11–12) *Tissue Engineering* 1658–1666.

[78] Xiaofeng Cui et al., 'Thermal Inkjet Printing in Tissue Engineering and Regenerative Medicine' (2012) 6(2) *Recent Patents on Drug Delivery and Formulation* 149–155.

[79] Fatema Tuj Zohora and Abu Yousuf Mohammad Anwarul Azim, 'Inkjet Printing: An Emerging Technology for 3D Tissue or Organ Printing' (2014) 10(30) *European Scientific Journal.*

[80] Ibid.

[81] E. S. Bishop et al., '3-D Bioprinting Technologies in Tissue Engineering and Regenerative Medicine: Current and Future Trends' (2017) 4(4) *Genes and Diseases* 185–195.

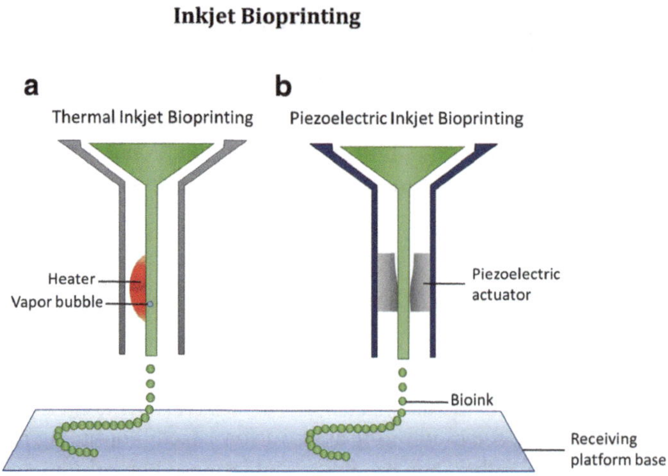

Fig. 2.5 Schematic visualisation of typed of inkjet bioprinting techniques. **a** Thermal inkjet bioprinting and **b** Piezoelectric bioprinting

printheads to deposit different cell types which is specifically useful in the fabrication of complex tissue structures.[82]

Most of the inkjet printers use thermal or mechanical compression to eject the drops of bioink. In the thermal inkjet process, minute air bubbles produced by heat in the printhead collapse to provide pressure pulses to eject bioink droplets with different volumes. The droplet sizes differ according to the temperature gradient applied, current pulse frequency, and viscosity of the bioink.[83] On the other hand, inkjet printers based on piezoelectric method print bioink on to the substrate through transient pressure received from the actuator of polycrystalline piezoelectric ceramic available in nozzles of the bioprinter.[84] Small bioprinters used for general small-scale output typically use thermal printers since they are less expensive to manufacture than piezoelectric bioprinters. However, piezoelectric bioprinters can deposit a wider variety of bioinks since they do not require a solvent with a low boiling point. It is for this reason that a majority of laboratory printers employ this bioprinting technique.[85]

Researchers in the field of bioprinting have been using inkjet methods in printing different molecules, stem cells, and tissues. Recently, thermal inkjet bioprinting techniques have been used to bioprint human microvascular endothelial cells. Endothelium refers to cells that line the interior surface of blood vessels and lymphatic

[82] Phil G. Campbell and Lee E. Weiss, 'Tissue Engineering with the Aid of Inkjet Printers' (2007) 7(8) *Expert Opinion on Biological Therapy* 1123–1127.

[83] Matthew E. Pepper et al., 'Cell Settling Effects on A Thermal Inkjet Bioprinter' 2011 Annual International Conference of the IEEE Engineering in Medicine and Biology Society.

[84] Jos de Jong et al., 'Air Entrapment in Piezo-Driven Inkjet Printheads' (2006) 120(3) *The Journal of the Acoustical Society of America* 1257–1265.

[85] Rachel Elizabeth Saunders and Brian Derby, 'Inkjet printing biomaterials for tissue engineering: bioprinting' (2014) 59(8) *International Materials Reviews* 430–448.

vessels, forming an interface between circulating blood or lymph in the lumen and the rest of the vessel wall.[86] In addition, scientists in collaborative research studies have designed a novel inkjet bioprinting method to create mechanically strong bone and cartilage tissue constructs.[87] In 2019, Saigopalakrishna and team used an inkjet-based bioprinting technique to construct spatially defined patterns of solid-phase exosome microenvironments. (Exosomes are a subset of extracellular vesicles that transport a wide variety of cell signalling molecules.) This first-of-its-kind research based on inkjet bioprinting offers the proof-of-concept for biofabricating extremely regulated and well-defined exosome-based microenvironments that can be used to study processes that support exosome mediated impacts on the extracellular matrix.[88]

Despite being one of the most used bioprinting techniques, inkjet-based bioprinting has its own limitations such as quality of vertical structure and low cell density. The bioprinter heads used in inkjet bioprinting are based on a specific type of device, which sometimes delivers a comparatively minute deformation created by either piezoelectric or thermal actuation at the nozzle opening. Due to this, bioprinter heads cannot release biomaterials that are highly viscous and as a result cannot perform well, and sometimes become clogged when bioinks with high cell density are loaded.[89] In addition, a study conducted by researchers at Clemson University emphasised another drawback of inkjet printing.[90] Bioinks are usually well mixed before loading into the bioink cartridge. As the printing process is time-consuming, the cells in the bioink tend to settle down in the cartridge and this sometimes leads to clogging of the printer head as a result of increased viscosity of the bioink. Thus, scientists are attempting to develop new advancements in inkjet bioprinting that can overcome these limitations and can be used to bioprint desired biomaterials in an efficient manner.

---

[86] L. H. Solis et al., 'Thermal Inkjet Bioprinting Triggers the Activation of the VEGF Pathway in Human Microvascular Endothelial Cells In vitro' (2019) 11(4) (2019/06/01) *Biofabrication* 045005.

[87] G. F. Gao et al., 'Improved Properties of Bone and Cartilage Tissue from 3D Inkjet-Bioprinted Human Mesenchymal Stem Cells by Simultaneous Deposition and Photocrosslinking in PEG-GelMA' (2015) 37(11) *Biotechnology Letters* 2349–2355.

[88] Saigopalakrishna S. Yerneni et al., 'Bioprinting Exosome-like Extracellular Vesicle Microenvironments' (2019) 13 *Bioprinting* e00041.

[89] T. Xu et al., 'Inkjet Printing of Viable Mammalian Cells' (2005) 26(1) *Biomaterials* 93–99; M. E. Pepper et al., 'Characterizing the Effects of Cell Settling on Bioprinter Output' (2012) 4(1) *Biofabrication*; M. E. Pepper et al., 'Cell Settling Effects on a Thermal Inkjet Bioprinter', *2011 Annual International Conference of the Ieee Engineering in Medicine and Biology Society* (2011) 3609–3612; B. Guillotin et al., 'Laser Assisted Bioprinting of Engineered Tissue with High Cell Density and Microscale Organization' (2010) 31(28) *Biomaterials* 7250–7256.

[90] Ibid.

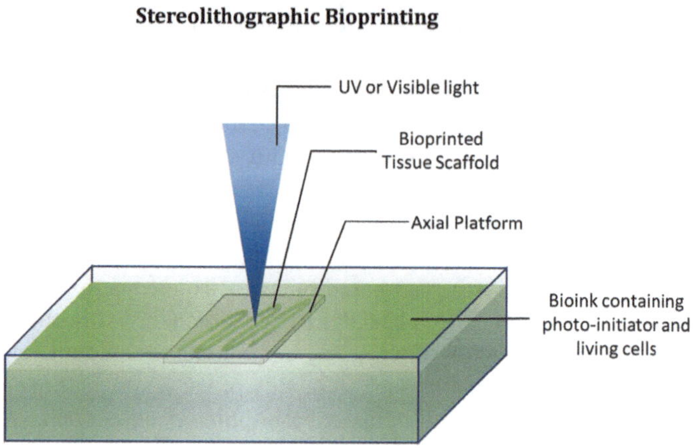

Fig. 2.6 Schematic visualisation of stereolithography-based bioprinting. Photopolymerisation occurs on the surface of the axial platform where the light-sensitive bioink is exposed to light energy to form layer-by-layer bioprinted tissue scaffold

2.5.3 Stereolithographic Bioprinting

Stereolithographic bioprinting technique works on the process of light-initiated polymerisation to bioprint 3D tissue scaffolds (Fig. 2.6). Generally, the process of stereolithographic bioprinting can be classified into two approaches: single photon and multi photon. The single photon technique can be further subclassified into: (i) Visible radiation techniques that use light in a visible wavelength range for polymerisation; (ii) Conventional stereolithographic techniques that use ultraviolet (UV) radiation for polymerisation; (iii) Infrared (IR) stereolithographic techniques that use radiation for polymerisation; and (iv) Stereo-thermal lithographic techniques that use UV as well as IR radiation during initiation of polymerisation.[91] Conventional single photon stereolithographic systems are most commonly used for bioprinting as this technique has the capability to process polymerisation with less heat creation that leads to minimal damage to the printed biomaterials and cells.[92] In comparison with stereolithographic techniques that use UV radiation for polymerisation, IR stereolithographic bioprinting needs advanced apparatus, which drives up equipment cost.[93]

Similar to the other bioprinting techniques, the steps involved in developing a 3D construct using stereolithography involve creating a blueprint of the desired tissue or organ using MRI or CT and developing a CAD file based on the blueprint. Based on

---

[91] Paulo Jorge Bártolo, 'Stereolithographic Processes', *Stereolithography* (Springer, 2011) 1–36.

[92] Ritu Raman and Rashid Bashir, 'Stereolithographic 3D Bioprinting for Biomedical Applications', *Essentials of 3D Biofabrication and Translation* (Elsevier, 2015) 89–121.

[93] Hitendra Kumar and Keekyoung Kim, 'Stereolithography 3D bioprinting', *3D Bioprinting* (Springer, 2020) 93–108. Ch 6.

the data available in the CAD file, the bioprinter creates the desired structure. Stereolithography specifically utilises digital micromirror arrays (an array of micromirrors primarily used for high speed, efficient, and reliable spatial light modulation) to regulate the light intensity in order to polymerise biomaterials that are light sensitive.[94] The advancement of biocompatible, biodegradable, and bioactive photocurable resins has allowed a wide range of biomedical and translational medical applications.[95] In comparison with the other solid freeform bioprinting methods, stereolithography has the greatest printing precision and can be used with different biomaterials. In addition, stereolithography has a unique ability to print biomaterials and hydrogels that are light sensitive.[96]

Recently, scientists at the University of California developed a stereolithography-based bioprinting platform for multi-material fabrication of heterogeneous hydrogel constructs. (A hydrogel is a network of polymer chains that are hydrophilic, sometimes found as a colloidal gel in which water is the dispersion medium.)[97] This bioprinting platform has an ability to rapidly change between different bioinks and can quickly accomplish multi-material layer-by-layer bioprinting.[98] In comparison with the conventional stereolithography-based bioprinters, this new bioprinting platform provides the benefit of multi-material fabrication at high resolution.[99] In 2018, researchers from the George Washington University developed a new multi-responsive structure using 4D stereolithography bioprinting. In this study, researchers demonstrated, 'a proof-of-concept intelligent nerve guidance conduit on a graphene hybrid 4D construct that offers multifunctional characteristics for nerve regeneration.'[100] Recent research conducted by German researchers has revealed an in vitro chondrocyte niche by investigation of different extracellular matrices.[101] Chondrocytes are the only cells found in healthy cartilage. They produce and maintain the cartilaginous matrix, which consists mainly of collagen and proteoglycans. By employing autologous biomaterial and modifying the structure of the bioprinted structures with precise patient-specific personalisation, this cutting-edge technique could be used in the creation of 3D cartilage models and in the treatment of articular cartilage defects.[102] Similarly, a cell benign approach was reported by researchers

[94] Ziyun Zhuang and Ho Pui Ho, 'Application of digital micromirror devices (DMD) in biomedical instruments' (2020) 13(06) *Journal of Innovative Optical Health Sciences* 2030011.

[95] Raman and Bashir (n 92).

[96] A. P. Zhang et al., 'Rapid Fabrication of Complex 3D Extracellular Microenvironments by Dynamic Optical Projection Stereolithography' (2012) 24(31) *Advanced Materials* 4266–+.

[97] A. K. Miri et al., 'Microfluidics-Enabled Multimaterial Maskless Stereolithographic Bioprinting' (2018) 30(27) *Advanced Materials*.

[98] Ibid.

[99] Ibid.

[100] S. D. Miao et al., 'Stereolithographic 4D Bioprinting of Multiresponsive Architectures for Neura Engineering' (2018) 2(9) *Advanced Biosystems*.

[101] T. Lam et al., 'Photopolymerizable Gelatin and Hyaluronic Acid for Stereolithographic 3D Bioprinting of Tissue-Engineered Cartilage' (2019) (2019/03/13) *The Journal of Biomedical Materials Research*.

[102] Ibid.

from Harvard Medical School.[103] This technology uses an aqueous two-phase emulsion bioink and directly bioprints a hydrogel construct which is porously structured.[104] This unique bioink is used to bioprint photo-crosslinked predesigned cell-laden hydrogel structures using stereolithographic bioprinting or extrusion-based bioprinting techniques.[105]

On the other hand, there are a number of drawbacks in relation to use of stereolithographic techniques for bioprinting, including: an absence of adequate biodegradable and biocompatible polymers; an inability to completely eliminate the supportive framework; a lack of capacity to create lateral gradients in the bioprinted structures; and other detrimental effects caused by residual photocuring reagents.[106] Mostly, stereolithographic bioprinting uses UV light to cure different 3D constructs. Sometimes, this may lead to harmful effects on the cellular components of bioprinted tissue due to its exposure to UV light.[107] Recently, researchers have developed a visible light-curable and visibly transparent resin for stereolithographic 3D printing, which also retains UV curability.[108] However, currently the use of this invention for 3D bioprinting is yet to be explored as in its current state it is only useful for 3D printing.

2.5.4 *Laser-Assisted Bioprinting*

Laser-assisted bioprinting technique was initially developed in 2002 and patented[109] by inventors from the United States Naval Research Laboratory. Since then, this technique has been applied in tissue engineering research and gradually emerged to bioprint 3D tissue constructs.[110] As visualised in Fig. 2.7, laser-assisted bioprinting works by depositing bioink onto a substrate with the assistance of a laser as the source of energy. The laser-assisted bioprinting system involves a bioink cartridge and substrate base on which the bioink is 3D printed. The bioink cartridge further comprises a glass slide, metal layer, and bioink. The substrate base consists of hydrogel and glass slide. The process of laser-assisted bioprinting initiates by focusing a laser pulse on the metal layer present in the bioink cartridge. Due to

---

[103] G. L. Ying et al., 'Aqueous Two-Phase Emulsion Bioink-Enabled 3D Bioprinting of Porous Hydrogels' (2018) 30(50) *Advanced Materials.*

[104] Ibid.

[105] Ibid.

[106] F. P. W. Melchels, J. Feijen and D. W. Grijpma, 'A Review on Stereolithography and its Applications in Biomedical Engineering' (2010) 31(24) *Biomaterials* 6121–6130.

[107] Zizhuo Zheng et al., 'Visible light-induced 3D bioprinting technologies and corresponding bioink materials for tissue engineering: a review' (2021) 7(7) *Engineering* 966–978.

[108] H. K. Park et al., 'A Visible Light-Curable yet Visible Wavelength-Transparent Resin for Stereolithography 3D Printing' (2018) 10 *Nature Publishing Group Asia Materials.*

[109] *US Patent No US6905738B*, filed on 08 February 2002 ('Expired' on 29 August 2019).

[110] Olivia Kérourédan et al., 'Laser-Assisted Bioprinting for Tissue Engineering', *Biomaterials and Nanotechnology for Tissue Engineering* (CRC Press, 2016) 287–304.

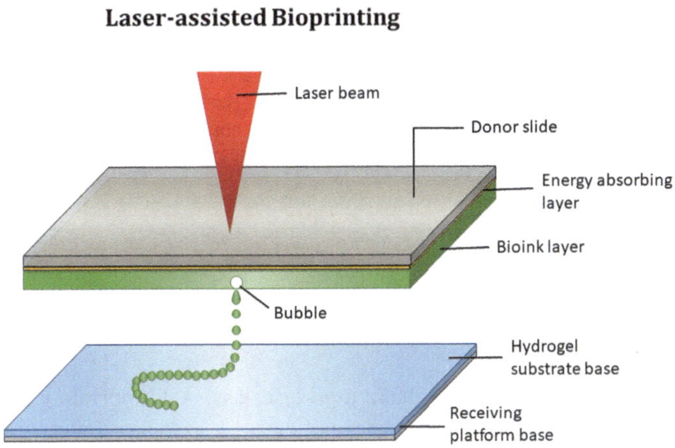

Fig. 2.7 Schematic visualisation of laser-assisted bioprinting. A typical laser-assisted bioprinting setup comprises a pulsed laser beam, a donor slide coated with an energy absorbing layer of metal, onto which a thin layer of bioink is spread, and a receiving platform base containing hydrogel substrate base

the laser pulse, bioink present on the metal layer is evaporated and deposited on to the substrate base. This process of printing one droplet of bioink on the substrate takes place in just 10 ms (milli seconds).[111] The steps described are continuously repeated as programmed to form a final 3D bioprinted construct. Initial versions of this technique have provided low cell visibility rates due to the direct contact of the laser to the living cells.[112] However, an upgraded version of laser-assisted bioprinting called 'laser-induced forward transfer' has been developed. This offers an additional 'energy absorption layer' which absorbs the energy released by the laser pulse and transfers a required amount of energy on to the slide containing the bioink.[113] Since the laser doesn't have direct contact with the living cells, this upgraded version of laser-assisted bioprinting has given rise to better cell viability rates.[114]

The laser-assisted bioprinting technique has been used to build advanced tissues constructs that can mimic native tissues. In a recent collaborative research study, scientists have developed human stem cell-based 3D corneal tissue structures with the help of a laser-assisted bioprinting technique.[115] In another study, French researchers

[111] Raphael Devillard et al., 'Cell Patterning by Laser-Assisted Bioprinting' in M. Piel and M. Thery (eds), *Micropatterning in Cell Biology, Pt A* (2014) vol 119, 159–174.

[112] D. J. Odde and M. J. Renn, 'Laser-Guided Direct Writing of Living Cells' (2000) 67(3) *Biotechnology and Bioengineering* 312–318.

[113] J. A. Barron et al., 'Application of Laser Printing to Mammalian Cells' (2004) 453 *Thin Solid Films* 383–387.

[114] B. Hopp et al., 'Survival and Proliferative Ability of Various Living Cell Types after Laser-Induced Forward Transfer' (2005) 11(11–12) *Tissue Engineering* 1817–1823.

[115] Anni Sorkio et al., 'Human Stem Cell Based Corneal Tissue Mimicking Structures using Laser-Assisted 3D Bioprinting and Functional Bioinks' (2018) 171 *Biomaterials* 57–71.

Table 2.1 A comparative overview of performance metrics of bioprinting methods[122]

Performance metrics	Extrusion	Inkjet	Stereolithography	Laser
Printing process	Serial (line by line)	Serial (drop by drop)	Parallel and continuous	Serial (dot by dot)
Cost	Medium	Low	Low	High
Cell viability	40–80%	>85%	>85%	>95%
Print speed	Slow	Fast	Fast	Medium
Resolution	Medium	High	High	High
Cell density	High	Low	Medium	Medium

from the University of Bordeaux have used laser-assisted bioprinting technique to successfully bioprint 'mesenchymal stromal cells,[116] associated with collagen and nano-hydroxyapatite,[117] in order to facilitate bone regeneration.'[118] In this manner, researchers are trying to utilise laser-assisted bioprinting techniques for the development of better tissue constructs that are able to mimic and interact with the native cellular environment.

In comparison with the other 3D bioprinting techniques, laser-assisted bioprinting offers improved automation, reproducibility, and high throughput.[119] Additionally, it overcomes some of the limitations of other 3D bioprinting techniques such as clogging of capillaries or printheads during inkjet bioprinting and the problems in printing high cell density droplets during extrusion-based bioprinting.[120] In contrast, the preparation of the bioink cartridge (glass slide, metal layer, bioink) is an additional time-consuming task, and sometimes the non-uniform coating of the bioink on the slide may lead to inaccurate cell output.[121] While each bioprinting technique has its own advantages and drawbacks, specific bioprinting techniques are employed considering the type of tissue that is required to be bioprinted. Table 2.1 further provides a comparative overview of the bioprinting methods discussed in this section.

[116] Mesenchymal stem cells are multipotent stromal cells that can differentiate into a variety of cell types, including osteoblasts (bone cells), chondrocytes (cartilage cells), myocytes (muscle cells), and adipocytes (fat cells which give rise to marrow adipose tissue).

[117] Nano hydroxyapatite is a mineral that can be used by the body to replenish lost enamel minerals and encourage new bone growth and development.

[118] Virginie Keriquel et al., 'In situ Printing of Mesenchymal Stromal Cells, by Laser-Assisted Bioprinting, for In vivo Bone Regeneration Applications' (2017) 7 *Scientific Reports.*

[119] F. Guillemot et al., 'High-Throughput Laser Printing of Cells and Biomaterials for Tissue Engineering' (2010) 6(7) *Acta Biomaterialia* 2494–2500.

[120] Fabien Guillemot et al., 'Laser-assisted bioprinting to deal with tissue complexity in regenerative medicine' (2011) 36(12) *Mrs Bulletin* 1015–1019.

[121] Nathan R. Schiele et al., 'Laser-Based Direct-Write Techniques for Cell Printing' (2010) 2(3) *Biofabrication.*

[122] Table designed using the relevant information provided in the following review article. Jie Xu et al., 'Advances in the research of bioinks based on natural collagen, polysaccharide and their derivatives for skin 3D bioprinting' (2020) 12(6) *Polymers* 1237. Table 2.1.

2.6 Conclusion

The historical context provided in the initial sections of this chapter discussed how bioprinting technology has evolved and how bioprinting technology differs from traditional tissue engineering and regenerative medicine. It is particularly important to comprehend the distinction between these technologies in order to understand how bioprinting technologies present new challenges to the interpretation of the patentable subject matter requirements in the context of patenting bioprinting inventions.

Recent advances in 3D bioprinting technology have accelerated the journey towards the development of complex bioinks that support composite vascular and functional bioprinted tissues. 3D bioprinting has progressed from single biomaterial or cell bioprinting to multi-material omnidirectional bioprinting, resulting in complex living tissues with built-in vasculature. Emerging 4D and in situ bioprinting technologies will increase bioprinting complexity by enabling the development of stimuli sensitive dynamic tissues. Transdisciplinary researchers have been continuously trying to develop different types of bioinks and methods to improve bio-printability, bioprinted tissue stability, vasculature, cell signalling, and growth. Despite the fact that bioprinting of fully functional human-scale 3D tissues and organs such as hearts, kidneys, and lungs requires more precise multicellular structures with vascular network integration, the recent advancements in bioprinting discussed in this chapter indicate that development of such functional 3D tissues and organs is progressing. The discussion on bioprinting technologies presented in this chapter is a window to assist in understanding the complexity of current bioprinting technologies and their impact on the scope of traditional patentable subject matter requirements analysed in further chapters of this book.

Chapter 3
3D Bioprinted Tissues and Bioinks: Falling in or out of the Scope of Subject Matter Requirements of Australian Patent Law

3.1 Introduction

Over centuries, patent systems around the globe have had to adapt to accommodate changing technologies. In the past few decades, however, the speed of technological advancement has been particularly demanding on global patent systems, requiring them to evolve and adapt to these advanced innovations more efficiently.[1] Research on 3D bioprinting technologies is a case in point. This research effort has escalated across the world in recent times.[2] Preliminary patent searches have indicated that most of the patent applications related to bioprinting are filed in the United States (US), Australia, and Europe.[3] Australia, Europe, and the US could thus be described as '3D bioprinting technology hubs,' justifying the choice of these three jurisdictions for analysis in this book.

While every jurisdiction has its own patent system, the Agreement on the Trade-related Aspects of Intellectual Property Rights (TRIPS) of the World Trade Organization (WTO) provides minimum standards which all member nations of the WTO are required to comply with.[4] Although member nations must maintain the obligations set by the Agreement, they have flexibility in how they apply these standards in their

---

[1] R. Levin, 'A Patent System for the 21st Century' (2004) 20(4) *Issues in Science and Technology* 49–54.

[2] 'A Bioprinting World Map' *3DPrint.com* (Web page, 19 June 2019) https://3dprint.com/246670/a-bioprinting-world-map/.

[3] Patent search results of patent applications filed in different jurisdictions on bioprinting as of 25 October 2019 https://www.lens.org/lens/search?q=bioprint*&l=en&st=true&preview=true&v=analysis.

[4] *Marrakesh Agreement Establishing the World Trade Organization*, opened for signature 15 April 1994, 1867 UNTS 3 (entered into force 1 January 1995) annex 1C (*Agreement on Trade-Related Aspects of Intellectual Property Rights*) ('*TRIPS*').

P. Devarapalli, *Bioinked Boundaries*, https://doi.org/10.1007/978-981-96-2591-8_3

domestic legal systems. Specifically, Article 27 of TRIPS provides the requirements for patentability.[5]

For inventions to be considered for patent grant, they must constitute patentable subject matter and must satisfy the core patentability criteria of novelty, inventive step/non-obviousness, utility/industrial applicability, and disclosure requirements.[6] Various jurisdictions around the world have further specific criteria in their patent laws on the inclusion or exclusion of particular types of subject matter to be considered for patent eligibility.[7]

3D bioprinting is one of the advanced technologies that now poses challenges to these traditional patent requirements.[8] For the purpose of this book, it is crucial to identify whether the specific subject matter of bioprinting technology falls within the scope of patentable subject matter and whether it is also likely to satisfy the other patentability criteria. This chapter examines patentable subject matter requirements in Australia in the context of 3D bioprinting innovations specifically with respect to products and processes related to bioinks and bioprinted tissues.[9]

As discussed in Sect. 2.3 of Chap. 2, bioinks are the materials that are used to develop living and functional tissues with the help of 3D bioprinting technology. The uniqueness of bioinks as subject matter is based on their composition, which is made up of a combination of naturally existing products and artificially created materials. The basic composition of bioinks includes naturally existing products such as living cells, bioactive molecules, and growth factors, as well as artificially created polymer gels that act as carriers and envelop the living cells, providing a structural configuration to the bioprinted cells.[10]. For example, a recent patent application filed by Oxford University with IP Australia claimed a bioink composition as:

> a composition comprising: (i) an aqueous medium, which aqueous medium comprises: (a) a hydrogel compound; (b) a culture medium; (c) one or more compounds comprising a dipeptide, and (ii) one or more biological cells disposed in the aqueous medium.[11]

In addition, the other class of subject matter this chapter focuses on is bioprinted tissues. Bioprinted tissues are constructs that are created using bioinks via specific

---

[5] Ibid., Article 27.

[6] Ibid., 27.1.

[7] Brieanna Dolmage, 'The Evolution of Patentable Subject Matter in the United States' (2005) 27 *Whittier Law Review* 1023.

[8] Seung-Schik Yoo, '3D-Printed Biological Organs: Medical Potential and Patenting Opportunity' (2015) 25(5) *Expert Opinion on Therapeutic Patents* 507–11; Phoebe H. Li, '3D bioprinting: Regulation, Innovation, and Patents', *3D Bioprinting for Reconstructive Surgery* (2018) 217–231. Lim, Pheh Hoon and Phoebe Li, 'Patentability of biofabricated human organs: products of nature or products derived from nature revisited', *Patenting Biotechnical Innovation* (Edward Elgar Publishing, 2022) 69–88.

[9] See Sect. 1.1 of Chap. 1.

[10] M. Hospodiuk et al., 'The Bioink: A Comprehensive Review on Bioprintable Materials' (2017) 35(2) *Biotechnology Advances* 217–239.

[11] *AU Patent No. AU2017225354A1*, filed on 28 February 2017 ('Published' on 13 September 2018).

bioprinting techniques.[12] These bioprinted tissues are capable of mimicking native tissue biochemical and mechanical properties.[13] Similar to bioink patents, patent claims directed towards bioprinted tissues comprise both product and method claims. For example, a recent patent application filed by Organovo with IP Australia claimed a 'living tissue construct' as:

> An engineered, living, three-dimensional connective tissue construct comprising: connective tissue cells cohered to one another to provide a living, three-dimensional connective tissue construct; wherein the construct is substantially free of pre-formed scaffold at the time of use.[14]

This chapter examines whether these types of bioprinting subject matter claims are likely to satisfy the patentable subject matter requirements of Australian patent law, and specifically whether they constitute 'manners of manufacture.'[15] The chapter further analyses precedents from relevant court cases delivered by Australian courts and the Australian Patent Office (APO) and examines how these decisions might impact the patentability of the above specified types of bioprinting subject matter in Australia The doctrinal analysis conducted in this chapter focuses on analysing Australian patent jurisprudence from a bioprinting viewpoint with a particular emphasis on patentable subject matter in relation to patent claims directed towards products, processes, and medical treatment methods. Considering the above, this chapter aims to answer the following research question of the book: 'Is the patentable subject matter requirement a hurdle to patenting bioprinting inventions, specifically bioinks and bioprinted tissues?'

3.2 Patentable Subject Matter in Australia

Australian patent legislation and patent jurisprudence stands as a governing instrument to assess whether the specific subject matter of bioprinting inventions is patentable. The Patents Act relies on the definition of 'manner of manufacture' derived from the Statute of Monopolies, and it includes considerations of novelty, usefulness, and non-obviousness. In contrast to the patent systems of some other jurisdictions such as Europe,[16] the Australian *Patents Act 1990* (Cth) (the Patents Act) does not provide any express exclusions of specific classes of subject matter that

---

[12] Li, Jipeng et al., 'Recent Advances in Bioprinting Techniques: Approaches, Applications and Future Prospects' (2016) 14(1) *Journal of translational medicine* 271.

[13] Z. Wang et al., '3D bioprinted functional and contractile cardiac tissue constructs' (2018) 70 *Acta Biomaterialia* 48–56.

[14] *AU Patent No. AU2017200162A1*, filed on 10 January 2017 ('Published' on 02 February 2017).

[15] Manner of manufacture is a process which is a proper subject matter of letters patent according to traditional principles: see *Patents Act*, s 18(1)(a).

[16] *European Patent Convention*, opened for signature 5 October 1973, 1065 UNTS 254 (entered into force 7 October 1977) Article 52(2).

are not to be considered patentable.[17] This section will describe the main criteria for patentable subject matter under the Patents Act and how they have been interpreted by the Australian courts in relation to bioprinting inventions. Specifically, s 18(1) of the Patents Act states that, 'A patentable invention is an invention that, so far as claimed in any claim: (a) is a manner of manufacture within the meaning of Section 6 of the Statute of Monopolies.' Section 6 of the Statute of Monopolies comes from old English legislation dating back to 1623. This section has been regarded as a flexible and adaptable test for patentability, which allows for the inclusion of new and emerging technologies such as bioprinting. Specifically, it states that:

> Provided also and be it declared and enacted that any declaration before mentioned shall not extend to any letters patent and grants of privilege, for the term of 14 years or under hereafter to be made of the sole working or making of any manner of new manufacture within this realm to the true and first inventor and inventors of such manufactures which others, at the time of making such letters or grant, shall not use, so as also they be not contrary to the law, nor mischievous to the state, by raising prices of commodities at home or hurt of trade or generally inconvenient.[18]

On this basis, for inventions to be considered for a patent they should be a 'manner of new manufacture.' The Australian judiciary, in different cases, has provided guidance on the assessment of inventions according to s 18(1) of the Patents Act. In 1959, the High Court of Australia (the 'High Court') determined the scope of the equivalent manner of manufacture requirement under the *Patents Act 1952* (Cth) in *National Research Development Corporation v Commissioner of Patents* (*NRDC*).[19] *NRDC* is the leading case on the interpretation of the manner of manufacture requirement. This case is considered as a watershed case in the history of Australian patent jurisprudence. The *NRDC* case provided boundaries on what subject matter should be considered as a patentable invention. According to the High Court, subject matter will be patentable where it consists of an artificially created state of affairs with economic significance. One of the primary reasons *NRDC* is considered to be a watershed decision is the idea that, instead of focusing on the section's language (s 6 of the 1623 *Statute of Monopolies*),[20] the assessment needs to be focused on the principles that have been used to apply the section when assessing a subject matter's patentability. This approach allows for a broad and dynamic interpretation of what constitutes a manner of manufacture, which can accommodate novel and inventive technologies such as bioprinting.

In this case, the issue for the court was to determine if a new process of weed eradication constituted a manner of manufacture and could therefore be considered patentable subject matter. The High Court dissected the manner of manufacture requirement and explained that the word 'manufacture' should not be interpreted in a way that restricts the scope of patentability but should be interpreted as a 'general

[17] Except the exclusion of human beings and biological processes for their generation from patentability: see *Patents Act*, s 18(2).

[18] 21 Jac 1 c 1 (1623).

[19] *National Research Development Corporation v Commissioner of Patents* (1959) 102 CLR 252.

[20] 21 Ja 1, c 3.

title' according to the *Statute of Monopolies* along with the developed principles of patent law.[21] Considering this, the right question that should be asked is: 'Is this a proper subject of letters patent according to the principles which have been developed for the application of s 6 of the Statute of Monopolies?'[22]

The decision further stressed that for the subject matter of an invention to be considered a manner of manufacture, it should belong to the useful arts, must provide a material advantage, and its value to the country is in the field of economic endeavour.[23] In addition, the court also held that to be patentable, subject matter must be of an 'artificially created state of affairs' that can provide a discernible effect.[24] The most significant takeaway from the *NRDC* decision is that there must be human action to produce a 'artificially created state of affairs' that has some discernible effect in order for a subject matter to be patentable. While the application of this criterion may vary on a case-by-case basis, it suggests that bioprinting inventions, involving human intervention and manipulation of biological materials, could potentially meet the manner of manufacture requirement if they also hold economic significance.

As detailed above, the *NRDC* decision has clarified the scope of patentable subject matter in Australia for specific subject matter. The *NRDC* decision has been considered as a benchmark for the assessment of patentable subject matter in different decisions by the APO, the Federal Court of Australia (FCA), the Full Court of the Federal Court of Australia (the FCFCA), and the High Court. Principles derived from the *NRDC* decision have assisted multiple decision makers in identifying the patentability of different classes of subject matter such as methods of medical treatment,[25] computer-related inventions,[26] and business methods.[27] Subsequent decisions delivered by the High Court have provided further guidance on how to determine whether specific subject matter is within the scope of patentable subject matter according to s 18(1) of the Patents Act.[28] An analysis of some of these key Australian case decisions will be provided in the upcoming sections of this chapter, with a focus on how they relate to subject matter related to bioprinted tissues and bioinks.

---

[21] 'The word 'manufacture' finds a place in the present Act, not as a word intended to reduce a question of patentability to a question of verbal interpretation, but simply as the general title found in the Statute of Monopolies for the whole category under which all grants of patents which may be made in accordance with the developed principles of patent law are to be subsumed'; *National Research Development Corporation* n (18) at p269.

[22] *National Research Development Corporation* n (18) at p269.

[23] Ibid., 22.

[24] Ibid., 25.

[25] *Joos v. Commissioner of Patents* [1972] HCA 38; *Anaesthetic supplies pty limited v. Rescare limited* [1994] 50 FCR 1; *Bristol-Myers Squibb Co v F H Faulding & Co Ltd* [2000] FCA 316.

[26] *International Business Machines Corporation v Commissioner of Patents* (1991) FCA 811; *CCOM Pty Ltd v Jiejing Pty Ltd* (1994) 51 FCR 260.

[27] *Grant v Commissioner of Patents* [2006] FCAFC 120.

[28] Brad Sherman, 'Before the High Court D'Arcy v Myriad Genetics Inc: Patenting Genes in Australia' (2015) 37(1) *Sydney Law Review* 135–146; William Van Caenegem, 'The Technicality Requirement, Patent Scope and Patentable Subject Matter in Australia' (2002) 13(1) *Australian Intellectual Property Journal* 309–327.

According to s 18(2) of the Patents Act, 'human beings and the biological processes for their generation' are excluded from patentability. The precise scope of this exclusion has been further explained in IP Australia's *Manual of Practice and Procedure* (the Manual). The Manual suggested that fertilised human ova and equivalents; zygotes, blastocysts, embryos, and foetuses; and totipotent human cells, including those cells that are the products of nuclear transfer procedures fall under the scope of s 18(2).[29] Although s 18(2) of the Patents Act requires that human beings and the biological processes for their generation are not patentable, this section may not affect the patentability of bioprinting technologies, as none of the subject matter of bioprinting is related to generation of a complete human being. The technology, in its current state, is only capable of bioprinting parts or tissues of the human body but not a complete human being.[30] As such, bioprinting inventions in their current state may not fall within the scope of s 18(2) of the Patents Act.

Regarding general inconvenience as provided in Section 6 of the Statute of Monopolies, there have been no known instance in which the condition that an innovation be not generally inconvenient was used as the main justification for invalidating a patent. As a result, there is no definitive guidance on when an invention may be considered generally inconvenient.[31] However, considering the applications of bioprinting technology in health and medicine, there may be arguments that can be made considering the 'general inconvenience' provision of Section 6 of Statute of Monopolies.[32] This provision may raise ethical and social concerns about the potential impact of bioprinting technologies on human dignity, identity, and welfare.

3.3 Analysis of Australian Patent Jurisprudence

The Australian judiciary has played a key role in determining how the Australian patent system applies to inventions in the fields such as biotechnology, information technology, and other advanced technologies by providing decisions in some landmark patent cases.[33] Patent applications are initially assessed by the Commissioner of Patents. Decisions on non-patentability of patent applications during normal examination can be re-examined by the Commissioner of Patents at the APO (or their

---

[29] IP Australia (2004), *Manual of Practice and Procedure,* November, Part 2.9.3.5.

[30] A. Atala and G. Forgacs, 'Three-Dimensional Bioprinting in Regenerative Medicine: Reality, Hype, and Future' (2019) 8(8) *Stem Cells Translational Medicine* 744–745; B. Grigoryan et al., 'Multivascular Networks and Functional Intravascular Topologies within Biocompatible Hydrogels' (2019) 364(6439) *Science* 458–+.

[31] IP Australia (2016), *Manual of Practice and Procedure,* November, Part 2.9.3.3.

[32] Chris Dent, "Generally inconvenient': the 1624 Statute of Monopolies as political compromise' (2009) 33(2) *Melbourne University Law Review* 415–453; 21 Jac 1 c 1 (1623).

[33] Sherman (n 28).

delegate). The Commissioner also hears pre-grant opposition proceedings.[34] Both levels of APO decisions can be appealed to the FCA. The FCA also makes first instance decisions with regard to revocation of patent claims post-grant. In either case, there is a right to appeal the decision to the FCFCA. On limited circumstances where there is an appealable error, decisions of the FCFCA can be appealed to the High Court. Decisions of the High Court have the highest precedential value as this Court is at the top of the hierarchy of Australian judicial system. It should be noted at this point that patent claims are commonly distinguished as product claims and method, or process claims.[35] Product claims are used to describe a thing or an item whereas method or process claims are used to describe a sequence of steps. This section will provide analysis of relevant judicial decisions and examine how specific precedents might impact patentability of products of bioprinting such as bioinks, bioprinted tissues, and methods of developing or delivering the same. It will also highlight some of the challenges and uncertainties that may arise from applying existing case law to novel and complex technologies such as bioprinting.

3.3.1 Analysis of Patent Jurisprudence Relating to Product Claims

3D bioprinting is an emerging technology and is giving rise to a variety of new subject matter in a short span of time.[36] As noted earlier, subject matter related to bioinks and bioprinted tissue constructs are the focus in this chapter as they are considered as the core aspects in the field of 3D bioprinting.[37] This section will analyse relevant judgements delivered by Australian courts and the APO and assess claims related to bioinks and bioprinted tissue constructs in light of case analysis. It will also discuss some of the implications and challenges that may arise from applying existing legal frameworks to innovative and disruptive technologies such as bioprinting.

3.3.1.1 Patenting Genetic Information as Products

Over the last two decades, patenting of genes or genetic information has been the subject of intense debate, particularly when the genetic information claimed

[34] Australian patent office provides pre-grant opposition system for standard patent applications, where any person may oppose the granting of a patent within three months of acceptance of the application being advertised in the official journal of patents.

[35] Alan C. Marco, Joshua D. Sarnoff and AW Charles, 'Patent claims and patent scope' (2019) 48(9) *Research Policy* 103790.

[36] S. D. Eswaramoorthy, S. Ramakrishna and S. N. Rath, 'Recent Advances in Three-Dimensional Bioprinting of Stem Cells' (2019) 13(6) *Journal of Tissue Engineering and Regenerative Medicine* 908–924.

[37] S. Ji, E. Almeida and M. Guvendiren, '3D Bioprinting of Complex Channels within Cell-Laden Hydrogels' (2019) 95 *Acta Biomaterialia* 214–224.

is unchanged from its natural form. Courts of different jurisdictions have provided different opinions on allowing or denying patent claims that are directed towards genetic information as products.[38] *D'Arcy v Myriad Genetics Inc (D'Arcy)*[39] *is a landmark precedent in Australian patent jurisprudence regarding patenting genes in Australia and their status as patentable subject matter.*[40] *In October 2015,* the High Court held that claims 1–3 of the Myriad patent[41] directed towards isolated nucleic acids coding for mutant or polymorphic forms of the BRCA1 polypeptide[42] did not meet the requirement of manner of manufacture and were inherently non-patentable under s 18(1)(a) of the Patents Act. This section will describe the background and the reasoning of the High Court's decision and how it may affect the patentability of bioprinting inventions that involve genetic information. The case was initially heard by the FCA,[43] with Justice Nicholas delivering judgement in the favour of the patent applicants (Myriad), explaining that,

> There is no doubt that naturally occurring DNA and RNA as they exist inside the cells of the human body cannot be the subject of a valid patent.[...]The disputed claims extend only to naturally occurring DNA and RNA which have been extracted from cells obtained from the human body and purged of other biological materials with which they were associated.[44]

In the following year in September 2014, an appeal was heard by the FCAFC sitting en banc, with five judges.[45] In this hearing, the judges analysed the claims in issue in light of the conclusions of *NRDC* and the US Supreme court's decision on the same patent[46] and again provided a judgement in favour of the respondents Myriad. Although the claims analysed in the *NRDC* case were directed towards methods, the FCFCA reasoned that its conclusions on the manner of manufacture requirements have broader scope in the context of product claims. Considering the principles set down in *NRDC*, the FCAFC reasoned that, 'the isolated nucleic acid, including cDNA,[47] has resulted in an artificially created state of affairs for economic benefit. The claimed product is properly the subject of letters patent. The claim is to an invention within the meaning of s 18(1) of the Act.' This reasoning suggests that

---

[38] *Association for Molecular Pathology v Myriad Genetics Inc* 186 L Ed 2d 124 (2013).

[39] *D'Arcy v Myriad Genetics Inc* [2015] HCA 35.

[40] J. C. Lai, 'D'Arcy v. Myriad Genetics: A Demand for the "Made" or "Non-Information" and Clear Subject Matter?' (2016) 47(5) *IIC-International Review of Intellectual Property and Competition Law* 537–568.

[41] *AU Patent No. AU686004B2*, filed on 11 August 1995 ('Expired' on 11 August 2015).

[42] Breast cancer type 1 (BRACA1) susceptibility protein is a protein that in humans is encoded by the BRCA1 gene.

[43] *Cancer Voices Australia v Myriad Genetics Inc* [2013] FCA 65.

[44] Ibid., 136.

[45] *D'Arcy v Myriad Genetics Inc* [2014] FCAFC 115.

[46] *Association for Molecular Pathology* (n 32).

[47] Complementary DNA (cDNA) is DNA that is produced from a single-stranded Ribonucleic Acid (RNA) template using the enzyme reverse transcriptase. In molecular biology, cDNA is also produced to examine the transcriptome patterns in tissue, single cells, or single nuclei in experiments like microarrays and RNA-seq.

bioprinting inventions that involve isolated genetic information may be considered as patentable subject matter, as they involve human intervention and manipulation to create a new state of affairs with economic value.

By this time, the first of the original applicants, Cancer Voices Australia, had become unincorporated. The second original applicant, Yvonne D'Arcy, took on the proceedings, appealing the decision of the FCAFC to the High Court. The decision of the High Court was handed down on 7 October 2015. The major question directed to the High Court in this case was whether the invention claimed a manner of manufacture and was, therefore, patentable subject matter under s 18(1)(a). Considering the previous watershed decisions such as *NRDC* and *Apotex Pty Ltd v Sanofi-Aventis Australia Pty Ltd*[48] and the relevant sections of the Patents Act, the bench of seven judges (French CJ, Kiefel, Bell, Gageler, Keane, Nettle, and Gordon JJ) answered this question in the negative. In this decision, three different judgements were handed down with four judges (French CJ, Kiefel, Bell, and Keane JJ) writing one judgement, two judges (Gageler and Nettle JJ) writing another judgement, and a single judgement (Gordon J). This decision overturned the previous judgements by the FCA and FCAFC and held that isolated genetic information is not a manner of manufacture because it does not involve sufficient human intervention or transformation to create a new state of affairs with economic significance. This decision may have significant implications for bioprinting inventions that involve genetic information, as they may be challenged on similar grounds as being mere discoveries rather than inventions.

In assessing the previous decision of the FCAFC, French CJ, Kiefel, Bell, and Keane JJ (the plurality) specifically identified that, 'the approach taken by the Full Court and urged by Myriad involves an apparently straightforward characterisation based on the formal terms of the patent identifying the isolated nucleic acids as products…'[49] In their judgement, the sequence of nucleotides was 'information' which translates to the same BRCA1 polypeptide sequence whether within the cellular environment or when isolated out of the human body. That same 'information' is an essential element of the claimed subject matter.[50] In addition to characterising the claimed isolated nucleic acid as information, their Honours also developed a new factorial approach to assess the patentability of claimed invention. Considering the conclusions of *NRDC*, their Honours outlined a number of factors that courts should consider when deciding whether the exclusive rights created by the grant of letters patent may be considered to be extendable to a specific class of claims under the provisions of Section 18(1)(a) of the Act.[51] Their Honours provided six factors, with the first two being directed towards the requirement derived from *NRDC* for an artificially created state of affairs with economic significance.[52] For the plurality, these first two factors apply exclusively when the claimed subject matter falls within the

[48] *Apotex Pty Ltd v Sanofi-Aventis Australia Pty Ltd* [2013] HCA 50.

[49] *D'Arcy v Myriad Genetics Inc* [2014] FCAFC 115.87.

[50] Ibid., 89.

[51] Ibid., 28.

[52] Ibid.

existing concept of manner of manufacture. The other four factors that were set out by their Honours include:

3. Whether patentability would be consistent with the purposes of the Act (particularly whether the invention with potentially negative effects on innovation; have a chilling effect on activities beyond those formally the subject of the exclusive patent rights; and the patentability would involve the court in assessing important and conflicting public and private interests and purposes)[53];
4. Whether the patentability of the invention would enhance or detract from the coherence of the law relating to inherent patentability[54];
5. Pertinently to Australia's place in the international community of nations and specific obligations under international law and the patent laws of other countries[55];
6. Whether patentability of invention would involve law-making of a kind which should be done by the legislature.[56]

The plurality explained that, these four 'other' factors apply when assessing the patentability of a class of subject matter that involves a significant new application or extension of the concept of manner of manufacture. In light of the above, their Honours concluded that, 'Although it may be said in a formal sense that the invention as claimed, referring to isolated nucleic acids, embodies a product created by human action, that is not sufficient to support its characterisation as a manner of manufacture.'[57]

Gageler and Nettle JJ, in examining the facts and evidence in light of *NRDC*, drew a clear line between the claims directed towards nucleic acid sequences as products and as methods. Their Honours explained that a method of using the BRCA1 sequences to identify the presence of cancerous cells in the human body can constitute patentable subject matter.[58] However, claims directed only towards BRCA1 sequences as products did not pass this threshold as they did not demonstrate an artificially created state of affairs of economic significance.[59] Accordingly, their Honours agreed with the orders proposed in the judgement of French CJ, Kiefel, Bell, and Keane JJ.[60]

Finally, Justice Gordon analysed the claims in question and held that,

> Myriad did not create, make, or alter those specific mutations and polymorphisms. Myriad identified that specific mutations and polymorphisms in the BRCA1 gene indicated a predisposition to breast cancer and ovarian cancer. That fact existed before Myriad worked it out.[61]

Considering the above, Gordon J also agreed with the orders proposed in the judgement of French CJ, Kiefel, Bell, and Keane JJ and held that 'the subject matter

---

[53] Ibid.

[54] Ibid.

[55] Ibid.

[56] Ibid.

[57] Ibid., 94.

[58] Ibid., 168.

[59] Ibid., 166–169.

[60] Ibid., 172.

[61] Ibid., 271.

of each of claims 1–3 is not patentable according to the principles which have been developed for the application of s 6 of the Statute of Monopolies.'[62]

The above analysis has described the background and the reasoning of the High Court's decision in D'Arcy and how it may affect the patentability of bioprinting inventions that involve genetic information. The High Court rejected the arguments of the FCA and FCAFC and held that isolated genetic information is not a manner of manufacture because it does not involve sufficient human intervention or transformation to create a new state of affairs with economic significance. The High Court also introduced a new factorial approach to assess the patentability of novel and complex technologies, which involves considering various factors such as consistency with the purposes of the Patents Act, coherence with the law, international obligations and standards, and legislative competence.

The factorial approach set out by the plurality in the *D'Arcy*'s decision has not been given a great deal of attention in the FCA and FCAFC decisions that immediately followed it. Specifically, in the *Commissioner of Patents v RPL Central*,[63] Justices Kenny, Bennett, and Nicholas held that claims directed towards 'a method and system for computerised collection of information' did not involve a new class of claim and hence it was unnecessary to examine the same in light of the *D'Arcy*'s factors.[64] Similarly, Justice Jagot in *Gilead Sciences Pty Ltd v Idenix Pharmaceuticals LLC*[65] found that the claims in issue were directed towards chemical compounds and pharmaceutical compositions which are not a new class of claim and thus do not require consideration of the *D'Arcy*'s factors.[66] Other decisions considering the assessment of patentable subject matter by the FCA and APO only discussed the first two *D'Arcy* factors directed towards 'human action' and 'economic utility' developed in light of the *NRDC* decision.[67]

Since the *D'Arcy* decision, the APO has been determining the patentability of related claims by applying two filters, one of which is 'identifying the substance of the claim' and the other is checking 'whether the substance of the claim has been made or changed by man, or is it artificial?'[68] IP Australia issued detailed instructions on the patentability of genetic information, including cDNA, in view of the principles provided by the High Court in D'Arcy. In particular, IP Australia's recommendations

---

[62] Ibid., 265.

[63] *Commissioner of Patents v RPL Central Pty Ltd* (2015) 238 FCR 27; [2015] FCAFC 177.

[64] Ibid., 119.

[65] *Gilead Sciences Pty Ltd v Idenix Pharmaceuticals LLC* (2016) 117 IPR 252; [2016] FCA 169.

[66] Ibid., 658.

[67] *Cargill Incorporated v Dow AgroSciences LLC* [2016] APO 43; *Rozenberg and Co Pty Ltd v Velin-Pharma A/S* [2017] APO 61; *InterPharma Pty Ltd v Hospira, Inc* (No 3) [2017] FCA 1536; *TPI Enterprises Ltd v Tasmanian Alkaloids Pty Ltd* [2018] APO 8; *Sun Pharmaceutical Industries (Australia) Pty Ltd v Tasmanian Alkaloids Pty Ltd* [2018] APO 7.

[68] Examination Practice following the High Court decision in *D'Arcy v Myriad Genetics Inc.* https://www.ipaustralia.gov.au/about-us/news-and-community/news/examination-practice-following-high-court-decision-darcy-v-myriad.

suggested that the act of isolation, purification, or synthesis is not enough to confer patentability of a genetic material including cDNA.[69]

In October 2016, a decision[70] by the APO involving a patent application[71] filed by Arrowhead Research Corporation (Arrowhead) determined that claims directed to double-stranded ribonucleic acid (dsRNA) molecules, which trigger RNA interference (RNAi), did constitute patentable subject matter. Ribonucleic acid (RNA) is a nucleic acid present in all living cells that has structural similarities to DNA. RNAi is a biological process in which RNA molecules inhibit gene expression or translation, by neutralising targeted messenger RNA molecules.

Initially, the examiner raised objections to claims 6, 7, and 9–13 explaining that they did not satisfy the requirement of manner of manufacture in light of the criteria set forth in the *D'Arcy* decision. Specifically, the examiner concluded that these claims were directed towards compositions of genetic information which were not 'made' as they were the same as in the naturally existing sequence of the genome[72] of the organism. Arrowhead responded to these objections and argued that, 'the substance of the invention was not genetic information but rather economically significant products and compositions.'[73] Further examining Arrowhead's arguments describing the substance of the invention as a whole and how it works, the Delegate of the Commissioner of Patents (the Delegate) noted that,

> When considering the invention as a whole, the dsRNA architecture of the claimed compositions is at least equally significant to the working of the invention, with the chemical and structural elements of the dsRNA being crucial for the recognition and digestion of the molecule by Dicer. As a consequence, the informational, structural, and chemical content of the dsRNA molecules may all be considered essential elements of the invention as claimed, and as a result the substance of the invention encompasses each of these elements, not merely the genetic information conveyed by the molecule.[74]

In light of the above, the Delegate identified the factors supporting the proposition that the substance of the invention is a pharmaceutical composition and not simply genetic information. The Delegate agreed with Arrowhead's submissions and concluded that the objection of manner of manufacture in relation to the present application had been overcome. In this way, multiple decisions of Australian courts and the APO have brought clarity on the patentability of claims directed towards genetic information or genes. Particularly, it is significant to note here that while the High Court's decision in *D'Arcy* has clearly indicated the non-patentability of an isolated gene sequence, later decisions related to patents with claims directed towards genetic information have tried to apply the 'human action' criteria to render a claim related to genetic information patentable. In light of the above analysis, the

[69] IP Australia (2020), *Manual of Practice and Procedure*, Part 2.9.2.6.

[70] *Arrowhead Research Corporation* [2016] APO 70.

[71] *AU Patent No. AU2013207601B2*, filed on 17 July 2013 ('Granted' on 17 November 2016).

[72] Genome is a complete set of genes or genetic material present in a cell or organism.

[73] *Arrowhead Research Corporation* [2016] APO 70. 20.

[74] Ibid., 29.

next section of this chapter will discuss how the above-discussed decisions are applicable to the patentability of products incorporating bioinks and bioprinted tissues. The above analysis has highlighted that while *D'Arcy* has set a clear precedent for denying patentability to isolated gene sequences, later decisions have tried to apply the 'human action' criterion derived from *NRDC* to render claims related to genetic information patentable.

3.3.1.2 Bioinks and Bioprinted Tissues as Products

Insights from the APO's decision considering Arrowhead's patent application are relevant to assess the patentability of product claims specifically directed towards bioink compositions. Specifically, if a bioink composition is composed of a genetic material (such as dsRNA), a claim directed towards such composition may be considered to satisfy the requirements of manner of manufacture. As previously discussed, bioinks are generally composed of living cells, growth factors, biopolymer gel, and other elements that are useful for the stabilisation and further proliferation of living cells.[75] Product and composition claims directed towards bioinks as a whole may be considered as economically significant products created by human action. In light of the above and the insights of this hearing, bioinks may be considered as compositions with 'economic significance' and may meet the manner of manufacture requirement. However, it is still too early to draw any conclusions about how the manner of manufacture requirements apply to inventions involving bioprinting and bioinks because of the ambiguity surrounding the applicability of these requirements as the application of which may vary on a case-to-case basis.

The High Court decision in *D'Arcy* also raises other questions on how product claims directed towards bioinks and bioprinted tissues would be considered patentable subject matter. Although the case does not provide any precise parameters to identify the level of human intervention needed for an invention to meet the requirements of manner of manufacture, the reasoning provided by the High Court may lead to different arguments when related to product claims directed towards bioinks and bioprinted tissues. The patentability outcome would differ on a case-by-case basis and the way that specific products claims are drafted. In light of the *D'Arcy* decision and its supporting judicial reasoning, the major concern in relation to consideration of bioprinting inventions as patentable subject matter is that inventions related to bioinks and bioprinted tissues use (or consist of) living cells.[76] For example, bioprinting inventions with product claims directed towards 'a bioprinted tissue'[77] may be rejected due to the reason that they are composed of 'living cells' and are not an artificially created state of affairs. However, the inventors can overcome this rejection with proper supporting evidence that specifies the extent of 'human

[75] See Sect. 2.3 of Chap. 2.

[76] *AU Patent No. AU2013277275B2*, filed on 19 June 2013 ('Granted' on 13 October 2016).

[77] *AU Patent No. AU2017200162A1* (n 14) Claim 1.

action' involved during the development of bioprinted tissue.[78] If a patent application is challenged by a patent examiner during the patent prosecution stage, asserting that it lacks an artificially created state of affairs, the applicant is provided with an opportunity to address this objection. The applicant can respond to the examiner by providing specific evidence or explanations regarding the extent of 'human action' involved in developing the claimed bioprinted tissue. For instance, the evidence may encompass details about the selection of an optimal bioink composition that enhances cell viability in the bioprinted tissue and the identification of precise parameters for bioprinting the tissue using a bioprinter.

Considering the four factors (*D'Arcy*'s factors 3–6) that were set out by French CJ, Kiefel, Bell, and Keane JJ in *D'Arcy* decision, it is important to know how these factors will influence the patentability of bioprinted tissues and bioinks. In particular, as suggested by their Honours, factors 3, 4, and 6 are of primary importance.[79] The plurality provided these factors in response to a broader concern considering the advanced and future technologies. Before looking into the individual factors in detail, it should be noted that these factors have been specifically developed for new classes of claims that involve a significant new application or extension of the concept of manner of manufacture.[80] Specifically, with reference to factor 3 sub factor 3.1 (whether the invention gives rise to a new field of monopoly protection with potentially negative effects on innovation), inventions claiming bioprinted tissues and bioinks might have the effect of boosting innovation in the field of bioprinting. In particular, the disclosure requirement could help other researchers in the field to gain new knowledge that would lead to further new innovations. With reference to sub factor 3.2 (whether the content of the claims might have a chilling effect on activities beyond those formally the subject of the exclusive rights granted to the patentee), claims directed towards bioprinted tissues and bioinks that only protect specified subject matter are unlikely to have any specific chilling effect on any activities beyond the patent rights. Considering sub factor 3.3 (conflicting public and private interests) with reference to bioprinted tissues and organs, it can be argued that bioprinting technology in general is closely related to the field of medicine, so there may be public interest concerns depending on how a specific invention has been claimed, which would vary on a case-to-case basis.

Similarly, factor 4 asks whether the patentability of the invention would enhance or detract from the coherence of the law relating to inherent patentability. Inherently patentable subject matter is defined as having characteristics that make specific subject matter inherently suitable for patent protection.[81] Factor 4 relates to the coherence of the law on inherent patentability, which requires that the subject matter be inherently suitable for patent protection. Bioprinting inventions may affect this factor depending on how they claim bioprinted tissues and bioinks. For instance, claiming

[78] *D'Arcy v Myriad Genetics Inc [2015] HCA 35. 136.*

[79] Ibid., 28.

[80] Ibid.

[81] Justine Pila, 'Inherent patentability in Anglo-Australian law: A history' (2003) 14 *Australian Intellectual Property Journal* 109–166.

only natural elements like stem cells may be seen as discoveries, not inventions, and thus inherently non-patentable. Concerning factor 6, which pertains to whether patenting would involve law-making that should be done by the legislature, the patentability of bioprinting inventions might require legislative intervention only if these inventions introduce a completely new type of subject matter. However, updated examination guidelines that outline patentable and non-patentable subject matter related to bioprinting inventions could suffice given the current state of bioprinting technology. It is recognised that the potential benefits associated with patenting in the field of bioprinting outlined above are highly speculative. While currently there is no compelling evidence to either support or refute these suggested benefits, the empirical component of this book will attempt to address some of them by providing specific patenting framework models.[82]

Contrary to the issue in *D'Arcy* (where the patent claims were directed towards isolated nucleic acids), developing a bioprinted structure is complex and involves multiple steps such as isolation, proliferation, and bioprinting.[83] According the *D'Arcy* reasoning,[84] the isolated living cells (by themselves) used for the development of bioinks and bioprinted tissues may not be patentable subject matter as they will be in the same natural state as in the human body and may be considered to be naturally existing products. However, the same isolated living cells, when further processed through precise growth conditions in combination with specific growth factors, bioactive molecules, and the biopolymer gel, providing an end product, may overcome some of the manner of manufacture requirements such as 'human action' and 'artificially created state of affairs' and may be considered as patentable subject matter under s 18(1)(a) of the Patents Act.

More particularly, in the *D'Arcy* case, claims 1–3 of Myriad's patent (as shown in Table 3.1) were directed towards isolated nucleic acid sequences as products which were not materially different to cellular nucleic acids naturally available in the cell. In contrast, claims directed towards 3D bioprinted tissue with living cells may note that a 3D bioprinted tissue construct is composed not only of living cells but also specific growth factors, bioactive molecules, and the biopolymer gel, which provides a protective environment for living cells to be active. A bioprinted tissue construct as a whole may also be considered to be an artificially created state of affairs as it is composed of different elements (as specified above) that are artificially brought together by the efforts of a human.

In light of the analysis provided in this section, it is evident that patent claims directed towards bioprinted tissues and bioinks as products may satisfy the requirements of the manner of manufacture test and may be considered patentable subject matter. However, due to the ambiguity inherent in the interpretation of subject matter requirements by different courts and variations in the types of bioprinting subject matter claimed in different patent applications, it means that decisions on

[82] See Sect. 7.2.1 of Chap. 7.

[83] P. Datta et al., 'Essential Steps in Bioprinting: From Pre- to Post-bioprinting' (2018) 36(5) *Biotechnology Advances* 1481–1504.

[84] *D'Arcy v Myriad Genetics Inc* [2015] HCA 35. *87*.

Table 3.1 A comparative overview of the first three claims of Myriad's patent[85]

Claim number	Claim description
1	An isolated nucleic acid coding for a mutant or polymorphic BRCA1 polypeptide, said nucleic acid containing in comparison to the BRCA 1 polypeptide encoding sequence set forth in 5 SEQ.ID No: 1 one or more mutations or polymorphisms selected from the mutations set forth in Tables 12, 12A and 14 and the polymorphisms set forth in Tables 18 and 19
2	An isolated nucleic acid as claimed in claim 1 which is a DNA coding for a mutant BRCA1 polypeptide, said DNA containing in comparison to the BRCA polypeptide encoding sequence 10 set forth in SEQ.ID No: 1 one or more mutations set forth in Tables 12, 12A and 14
3	An isolated nucleic acid as claimed in claim 1 which is a DNA coding for a polymorphic BRCAI polypeptide, said DNA containing in comparison to the BRCA1 polypeptide encoding sequence set forth in SEQ.ID No: 1 one or more polymorphisms set forth in Tables 18 and 19

patentability may vary on a case-to-case basis. The empirical phase of this book provides patent prosecution analysis of individual patent applications with claims directed towards bioprinted tissues and bioinks to provide insights on how patent offices are applying the judicial interpretations of these tests to assess specific bioprinting inventions.

3.3.2 *Analysis of Relevant Precedents Related to Method Claims*

Subject matter related to methods of developing and delivering bioinks and bioprinted tissues is of prime importance as they are comparatively complex due to the presence of living cells and artificially created elements. For example, a recent patent application filed by Organovo at IP Australia claims,

> A method for making an implantable tissue or organ comprising a muscle cell-containing layer, the method comprising: a) bioprinting bio-ink comprising muscle cells into a form; and b) fusing the bio-ink into a cohesive cellular structure; provided that the tissue or organ is implantable in a vertebrate subject and not a vascular tube.[86]

Numerous similar method claims have been filed with the APO, raising questions about their patentability based on relevant Australian court and APO decisions. This chapter focuses on assessing whether these methods are considered patentable subject matter, taking into account the guidance provided by *NRDC* and its application in Australian courts.

---

[85] *AU Patent No. AU686004B2*, filed on 11 August 1995 ('Expired' on 11 August 2015).

[86] *AU Patent No. AU2019204968A1*, filed on 11 July 2019 ('Published' on 01 August 2019).

3.3.2.1 How Have the Australian Courts Applied the Guidance of *NRDC*?

Post *NRDC*, Australian courts have analysed and applied the conclusions of the *NRDC* case to various subject matters. Justice Burchett in *International Business Machines Corporation v. Commissioner of Patents*[87] (IBM) considered the guidance of *NRDC*. In this case, the subject matter in question related to a method and apparatus for producing a curve image using algorithms. His Honours compared the facts of that case with the reasoning of *NRDC* and reasoned that,

> It seems to me that the use of the algorithm is not different conceptually from the use of the compounds involved in National Research and Development Council. Just as those compounds were previously known, so here, it is not suggested there is anything new about the mathematics of the invention.[88]

Justice Burchett identified that the application of the selected mathematical methods to computers, and in particular, to the production of the desired curve by a computer was inventive and commercially useful, which may imply that he considered it as a manner of manufacture.

Methods of Medical Treatment

In Australia, medical treatment methods are expressly not excluded from patentability under the Patents Act. Prior to 1972, Australian law considered surgical or medical treatment of the human body, as well as non-medical operations such as cosmetic treatment, to be patentable.[89] In 1972, Chief Justice Barwick in the High Court in *Joos v. Commissioner of Patents*[90] (*Joos*) reversed the decision of the Commissioner of Patents to refuse the grant of a patent for a process of treating human hair and nails. In his decision, his Honour related the 'economic significance' aspect of the *NRDC* case to the process for improving strength and elasticity of keratinous material and held that,

> The processes are to be used in what cannot be described otherwise than as a commercial activity of hairdressing, a sector of activity which accounts, I imagine, for a great deal of employment…it is an activity in the field of economic endeavour and has commercial significance as those expressions ought to be understood in relation to the grant of patents.[91]

By distinguishing cosmetic treatments from medical methods, Barwick CJ's decision paved the way for patenting processes that enhance or alter the appearance of the human body, opening up new possibilities for commercial applications.

---

[87] *International Business Machines Corporation v Commissioner of Patents* [1991] FCA 811.

[88] Ibid., 16.

[89] Australian Law Reform Commission, 'Genes and ingenuity: gene patenting and human health' (2004) *Commonwealth of Australia*. 7.29.

[90] *Joos v Commissioner of Patents* [1972] HCA 38.

[91] Ibid., 28.

The issue of patentability of methods of medical treatment was addressed in detail in the later years. In *Anaesthetic Supplies Pty Ltd v. Rescare Ltd* (1994) 50 FCR1 (*Rescare*), the FCFCA delved deeper into the patentability of medical treatment methods.[92] The majority of the court concluded that there was no justification for making a distinction between processes that result in a cosmetic or functional result as opposed to a curative result when determining patentability. In the current case, Heerey J followed the minority opposing judgement in *Rescare* because he considered the majority's conclusion to be essentially obiter dictum. The FCFCA later reversed this portion of the first instance judgement on appeal for two key reasons. First, Parliament chose not to exempt medical treatment methods from patentability under the *Patents Act*.[93] Second, the court found no logical justification for granting patents for pharmaceutical products but not for medical treatment methods that employ such products.[94]

Following from *NRDC* and the later decisions, in 2013 the High Court in *Apotex Pty Ltd v Sanofi-Aventis Australia Pty Ltd*[95] held that a method of preventing or treating psoriasis by administering leflunomide should be considered patentable subject matter within the meaning of s 18(1)(a) of the Patents Act. Australian courts had previously addressed patentability of medical treatment methods in various instances.[96] However, the appellant Apotex challenged the earlier Federal Court decisions, which raised significance as it involved a method of treatment and the administration of a medication dose. On the other hand, cases dealing with other forms of treatment typically focused more on surgical methods.

The issue arose in 2008 when Sanofi commenced proceedings against Apotex claiming that Apotex's proposed supply of leflunomide, a compound that is used to treat psoriatic arthritis (PsA), would infringe its patent.[97] Apotex cross-claimed seeking revocation of Sanofi's patent on a number of grounds, including that a method of medical treatment of the human body is not patentable subject matter under Section 18(1) of the Patents Act. At first instance, Justice Jagot analysed the facts and arguments, in light of *NRDC* and later decisions such as *Anaesthetic Supplies Ltd v Rescare Ltd* (*Anaesthetic*)[98] and *Bristol-Myers Squibb Co v FH Faulding & Co* (*Bristol*)[99] with reference to the 'requirement of newness' (if it is apparent on the face of the specification that the claimed invention is not new) and concluded that Apotex's argument lacked the necessary quality of newness inherent in a manner of manufacture as required by s 18(1)(a) of the *Patents Act*.[100]

---

92 *Anaesthetic Supplies Pty Ltd v Rescare Ltd* [1994] 50 FCR 1.

93 Ibid., 3.

94 Ibid., 26.

95 *Apotex Pty Ltd v Sanofi-Aventis Australia Pty Ltd* [2013] HCA 50.

96 *Anaesthetic Supplies Pty Ltd v Rescare Ltd* [1994] 50 FCR 1; *Bristol-Myers Squibb Co v FH Faulding & Co Ltd* [2000] 170 ALR 439.

97 *AU Patent No. AU670491B2*, filed on 29 March 1994 ('Expired' on 29 March 2014).

98 *Anaesthetic Supplies Pty Ltd v. Rescare Ltd* [1994] FCA 304.

99 *Rescare Ltd v Anaesthetic Supplies Pty Ltd* [1992] FCA 811.

100 *Sanofi-Aventis Australia Pty Ltd v Apotex Pty Ltd* (No 3) [2011] FCA 846 240, 243.

Apotex further appealed the FCA decision to the FCFCA. The decision of the FCFCA was handed down in July 2012. The bench of three judges (Keane CJ, Bennett, and Yates JJ) analysed the case and considered the reasoning and conclusions in other methods of medical treatment cases, particularly focusing on the FCAFC decisions in *Anaesthetic*[101] and *Bristol.*[102] Keane CJ emphasised that Parliament did not intend the area of non-patentability associated with the preservation of human health to be wider than the express statement in s 18(2) of the *Patents Act.*[103]

Keane CJ further concluded that the patent claim, properly construed, does not extend to using leflunomide as a method of treating PsA merely because of its incidental beneficial effect on psoriasis.[104] Considering Keane CJ's views, the other two judges, Bennett and Yates JJ, heard both parties' arguments and held that they would not deviate from the position supporting the patentability of medical treatment methods, which they regarded as the established norm in Australian patent law. They emphasised that patents for methods of medical treatment are routinely granted in Australia.[105] The FCAFC upheld the trial judge's decision and rejected the submissions of Apotex. Apotex subsequently applied for special leave to appeal to the High Court. Special leave was granted in December 2012.[106]

The High Court (French CJ, Hayne, Crennan, Kiefel, and Gageler JJ) addressed the key issues of the case and the major question that a method of medical treatment be a manner of manufacture and thus patentable. Considering the importance of 'economic significance' for the assessment of 'manner of manufacture,' the High Court applied and reanalysed *NRDC*'s 'economic significance' principle in the context of how economically significant the relevant methods of medical treatment are. Chief Justice French emphasised that, 'In my opinion the application of the rubric "manner of new manufacture" in a logically and normatively coherent way is not served by excluding from its scope methods of medical treatment of human beings.'[107]

In contrast, Justice Hayne analysed the subject matter from a different perspective, emphasising the 'economic significance' aspect and suggested that the end product of a method of medical treatment is economically significant to the individual but not to the country or society. Justice Hayne opined that a 'method of prevention or treatment of human disease is not a proper subject for the grant of a patent.'[108]

However, Justices Crennan and Kiefel took a different perspective, emphasising that for patent law, the subject matter must have some useful application in commerce or industry.[109] Considering that 'methods of medical treatment of the human body'

---

[101] *Anaesthetic Supplies Pty Ltd v Rescare Ltd* [1994] 50 FCR 1.

[102] *Bristol-Myers Squibb Co v FH Faulding & Co Ltd* [2000] 170 ALR 439.

[103] *Apotex Pty Ltd v Sanofi-Aventis Australia Pty Ltd* (No 2) [2012] FCAFC 102 26.

[104] Ibid., 27.

[105] Ibid., 193.

[106] *Apotex Pty Ltd v Sanofi-Aventis Australia Pty Ltd & Ors* [2012] HCATrans 357.

[107] *Apotex Pty Ltd* (n 81) 50.

[108] Ibid., 165.

[109] Ibid., 278.

satisfy the 'economic significance' principle, Justices Crennan and Kiefel made a distinction between a method of medical treatment which involves an unknown therapeutic use of a pharmaceutical and the medical activities or procedures of doctors when physically treating patients.[110] Their Honours considered that claims directed towards a method of medical treatment or product that involves an unknown therapeutic use of a known pharmaceutical product have economic utility and also provide an artificially created improvement in the human body.[111] In contrast, their Honours considered that inventions claiming medical activities or procedures of doctors (and other medical staff) when physically treating patients are 'essentially non-economic' and further not 'susceptible' or 'capable' of industrial application.[112] The court acknowledged that there is a distinction between methods of medical treatment and the medical activities or procedures of doctors for the treatment of patients.

As noted earlier, s 18(2) of the *Patents Act* excludes human beings and biological processes for their generation from patentability in Australia.[113] In August 2016, the APO provided useful interpretation of s 18(2)[114] in a decision considering methods directed to the parthenogenic activation of a human oocyte for the production of human stem cells[115] as patentable subject matter.[116] The case concerned two patent applications filed by International Stem Cell Corporation (ISCC), with one patent, 2012216371 ('371'), including claims directed towards the production of synthetic cornea from stem cells[117] and the other, 2013205483 ('483'), including claims directed towards parthenogenic activation of human oocytes for production of human stem cells.[118]

Initially, examiners rejected both patent applications, arguing that the claims did not constitute patentable subject matter according to s 18(2) of the *Patents Act*, as they were seen as a biological process for the generation of a human being.[119] However, the applicants responded to the objections, providing evidence that the blastocyst[120] formed via parthenogenic activation of unfertilised human oocytes could not lead to

---

[110] Ibid., 287.

[111] Ibid., 282.

[112] Ibid., 287.

[113] *Patents Act 1990* (Cth) s 18(3); Sect. 3.2 of this chapter.

[114] Ibid.

[115] Parthenogenesis is a natural form of asexual reproduction in which growth and development of embryos occur without fertilisation. Parthenogenetic activation is used for biochemical and morphological studies of oocytes during fertilisation and early development and is a critical component of the cloning procedure; Matthew H. Kaufman, 'Parthenogenetic Activation of Oocytes' (2018) 2018(1) *Cold Spring Harbor Protocols*; An oocyte is a cell in an ovary that multiplies to form an ovum; A stem cell is a living cell in the human body that has a special ability to develop into specialised cell types such as muscle cells, brain cells, etc.

[116] *International Stem Cell Corporation* [2016] APO 52.

[117] *AU Patent No. AU2012216371A1*, filed on 15 August 2012 ('Published' on 13 September 2012).

[118] *AU Patent No. AU2017201485B2*, filed on 29 April 2013 ('Granted' on 29 September 2016).

[119] *International Stem Cell Corporation* [2016] APO 52. 16.

[120] Blastocysts are structures formed in the early development stages of mammals which subsequently transform into an embryo.

further stages of birth and thus could not give rise to a human being.[121] Based on this, the Delegate concluded that the claims did not fall within the scope of 'biological processes for the generation of human beings'[122] and were, therefore, patentable subject matter.[123]

This APO decision offers guidance on applying s 18(2) to subject matter related to biological processes for the generation of a human being. As discussed earlier, subject matter related to methods of developing bioinks and bioprinted tissues may not fall within the scope of the s 18(2) exclusion. Here, it is important to identify how specific claims directed towards methods of using bioinks and bioprinted tissues would be considered an 'artificially created state of affairs' as their makeup is a combination of synthetic elements and naturally existing products such as living cells, growth factors, and biopolymer gel. Noting the judgement of Crennan and Kiefel JJ in *Apotex*, methods of using bioinks and bioprinted tissues related to treatment of human beings that include claims directed towards physical treatment of patients by medical practitioners may not be patentable subject matter within the meaning of s 18(1)(a) of the Patents Act. However, considering the importance of 'economic significance' for the assessment of 'manner of manufacture,' the claims directed towards methods involving bioinks and bioprinted tissues have applications in pharmaceutical, health, and medicine industries. Specifically, such inventions are directly or indirectly useful for the treatment of human body and can considered 'economically significant.'[124] For example, a method of treating a disease in a tissue or organ can be considered as a method of medical treatment by application of a pharmaceutical and may be considered patentable subject matter.[125] Therefore, some of the related claims directed towards 'medical treatment' may be patentable subject matter. However, based on the judgement of Crennan and Kiefel JJ in *Apotex*, it is advisable that inventors draft their claims carefully so that they do not claim the surgical process or methods of using bioinks and bioprinted tissues involving doctors as these claims may be considered as medical activities or procedures of doctors (and other medical staff) physically treating the patients, and further will be considered 'essentially non-economic.' Given the pace of advancement of bioprinting and its relevance to the development of human body parts, the decisions discussed in this section (related to methods of medical treatment) provide useful guidance to assess the patentability of relevant bioprinting subject matter in the near future.

---

[121] *International Stem Cell Corporation* [2016] APO 52. 29.

[122] Ibid., 30.

[123] Ibid., 32.

[124] Jamroz et al. (n 64); Gao et al. (n 28).

[125] *WIPO Patent Number WO2019122351A1*, filed on 21 December 2018 ('Published' on 27 June 2019).

3.3.2.2 Methods Claiming Practical Application of Naturally Existing Products

After the High Court's decision in D'Arcy, the APO updated the Manual[126] to reflect the conclusions. The Manual[127] provides guidance on patentability of product and method claims involving naturally existing products and specifically suggests that claims directed towards methods using naturally existing products would be considered patentable subject matter.[128] However, in 2018, Meat & Livestock Australia (MLA) sued in the FCA[129] against a patent application (Branhaven patent)[130] filed by Cargill, Inc and Branhaven LLC. The patent relates to methods identifying genetic differences between cattle and associating them with specific traits to select and breed cattle with preferred traits. MLA had previously opposed the patent at the APO, and the Delegate held all claims valid (except some amendments to claim 13), prompting MLA to appeal to the FCA. Some of the disputed claims in the Branhaven patent involve Single Nucleotide Polymorphisms (SNPs)[131] as biological markers. MLA argued that these SNPs exist in nature and are not patentable subject matter under the *Patents Act*, as interpreted by the High Court in *D'Arcy*.[132]

It is important to recall that in the *D'Arcy* case, the claims focused on product claims concerning nucleic acid sequences. Justice Beach distinguished *Meat & Livestock Australia Limited v Cargill* (*Cargill*)[133] from *D'Arcy*[134] based on the Branhaven patent's claims. Unlike *D'Arcy*, the Branhaven patent's claims (except claim 13)[135] were directed towards methods rather than products. Specifically, claim 1 of the Branhaven patent is directed towards a method for identifying a trait of a bovine subject using Single Nucleotide Polymorphisms (SNPs) present in a nucleic acid sample.[136]

Analysing claim 1 of the Branhaven patent, Justice Beach concluded that it involved the practical application of naturally existing products (SNPs) for a specific purpose, namely the process of trait identification in cattle. He emphasised that the claimed invention was not merely identifying or discerning a naturally occurring phenomenon, contrary to MLA's assertions. Instead, the method required human

---

[126] IP Australia (2016), *Manual of Practice and Procedure,* November, Part 2.9.1.

[127] Ibid. 2.9.2.6.

[128] Ibid.

[129] *Meat & Livestock Australia Limited v Cargill, Inc* [2018] FCA 51.

[130] *AU Patent No. AU2010202253B2*, filed on 01 June 2010 ('Granted' on 31 October 2013).

[131] A Single Nucleotide Polymorphism is a variation of DNA sequence which occurs when a single nucleotide in the genome differs between members of a species or paired chromosomes in an individual.

[132] *Meat & Livestock Australia Limited v Cargill, Inc* [2018] FCA 51. 13.

[133] Ibid.

[134] *D'Arcy v Myriad Genetics Inc* [2015] HCA 35.

[135] *AU Patent No. AU2010202253B2* (n 123) Claim 13; An isolated polynucleotide identified according to the method of claim 8.

[136] *AU Patent No. AU2010202253B2* (n 123) Claim 1.

interaction and the creation of an artificially created state of affairs. The process entailed taking a nucleic acid sample from the bovine subject, identifying at least three SNPs associated with the trait, and analysing the sample to meet the claim's requirements. This method resulted in an economically significant artificially created state of affairs through human interaction.[137]

Although Justice Beach's conclusions in this case are reasonably justified with reference to previous court decisions such as *NRDC* and *D'Arcy*, MLA appealed the decision. The decision of the FCAFC was handed down in 2020.[138] However, the appeal was mostly concerned with patent amendments as provided by Sections 105(1A) and 112A of the Patents Act and was not related to patentable subject matter concerns.[139]

Justice Beach delivered another decision in June 2019 in *Sequenom, Inc. v Ariosa Diagnostics, Inc.* (*Sequenom*) considering whether diagnostic methods are patentable subject matter.[140] In March 1998, Sequenom, Inc. filed a patent application[141] related to 'Non-invasive prenatal diagnosis' and after examination, IP Australia granted the patent application. One of the claims of the patent in question was directed towards, 'A detection method performed on a maternal serum or plasma sample from a pregnant female, which method comprises detecting the presence of a nucleic acid of foetal origin in the sample.'[142]

Sequenom, Inc's patented technology works by detecting cell-free foetal DNA (cffDNA) (foetal DNA which circulates freely in the maternal blood) in maternal blood serum and blood plasma. This technology is useful for prenatal diagnosis of a range of conditions, including foetal abnormalities, blood typing, and sex determination.[143] Sequenom, Inc. brought infringement proceedings in the FCA against Ariosa Diagnostics, Inc., and others (Respondents) for alleged use of the patented technology in the Australian market.[144] In response to this, Ariosa cross-claimed for revocation of the patent on the basis that the relevant claims did not define patentable subject matter.

In particular, the respondents argued, in light of *D'Arcy*, that claims directed towards cffDNA did not qualify as a 'manner of manufacture.' They asserted that the invention merely discovered cffDNA, making the end result genetic information of the foetal DNA, and not an 'artificially created state of affairs.' Justice Beach, who also presided over *Meat & Livestock Australia Limited v Cargill, Inc.*, rejected

[137] *Meat & Livestock Australia Limited v Cargill, Inc* [2018] FCA 51. 455.

[138] *Meat and Livestock Australia Limited v Branhaven LLC* [2020] FCAFC 171.

[139] Ibid.

[140] *Sequenom, Inc. v Ariosa Diagnostics, Inc.* [2019] FCA 1011.

[141] *AU Patent No. AU727919B2*, filed on 04 March 1998 ('Granted' on 19 April 2001).

[142] Ibid., Claim 1.

[143] Ibid., Claim 12, 13.

[144] San Jose, 'Ariosa Diagnostics' Harmony™ Prenatal Test now performed in Australia and United Kingdom at Sonic Healthcare laboratories', *Roche* (Webpage, 14 November 2015) https://sequencing.roche.com/en/news-overview/2015/ariosa-diagnostics-harmony-prenatal-test-now-performed-in-austra.html.

these arguments. He clarified that the patent did not merely claim the discovery of cffDNA in maternal blood; instead, it presented a new and inventive practical application of the discovery. The method required human action to detect a specific DNA sequence as being of foetal origin in an artificially created sample of maternal plasma or serum.[145]

His Honour focused particularly on how the principles must be applied differently for product and method claims. Justice Beach indicated that Sequenom, Inc.'s claims are directed towards a method of practical application of a natural phenomenon rather than directly claiming a natural phenomenon, namely the presence of cffDNA in maternal blood and not the cffDNA itself.[146] Addressing the issue of 'human intervention,' Justice Beach reasoned that cffDNA detection in a sample resulted from this process. Without human action, cffDNA could not be detected in maternal blood. This contrasted with the naturally occurring nucleotides in Myriad or the genetic information encoded by sequences of cffDNA, as the physical detection result in this case was achieved through human action.[147]

The main essence of Justice Beach's analysis is that he distinguished this case from *D'Arcy*, where Myriad's claims were directed towards DNA itself as a product lacking the required 'human action.' In contrast, Sequenom, Inc.'s patent claims were focused on a 'method' of 'detecting' cffDNA, which inherently involved 'human action.'

Ariosa Diagnostics Inc. appealed Justice Beach's decision, with the resulting decision being reported as *Ariosa Diagnostics Inc v Sequenom Inc* (*Ariosa*).[148] The central problem on appeal, as acknowledged by the Full Court, was whether what was claimed was a method having practical application that extends beyond the discovery itself, or merely a finding of a found natural phenomenon.[149] Ariosa contended that the claim was merely the discovery of a naturally occurring phenomenon (information about the foetal DNA),[150] and so there was 'no artificially created state of affairs,' as required by the manner of manufacture test.[151] Ariosa attempted to draw an analogy with the High court's analysis in *D'Arcy*.[152]

Drawing a distinction between naturally occurring phenomena and their practical applications was key to the Full Court's analysis in this case. Accordingly, the Full Court stated that:

> the distinction between mere discovery and an invention lies in its practical application to a useful end … an invention may reside in an abstract idea, such as the condensation of steam,

[145] *Sequenom, Inc. v Ariosa Diagnostics, Inc.* [2019] FCA 1011. 526.

[146] Ibid., 485.

[147] Ibid., 496.

[148] *Ariosa Diagnostics Inc v Sequenom Inc* [2021] FCAFC 101.

[149] Ibid., 108.

[150] Ibid., 95.

[151] Ibid., 98.

[152] *D'Arcy v Myriad Genetics Inc* [2015] HCA 35.

> that is then put to a useful end, even though the way of putting it to that end can be carried out in many useful ways, all of which are otherwise known.[153]

The Full Court rejected Ariosa's arguments, emphasising that the facts in this case were fundamentally different from those in Myriad. Particularly considering that the claim at issue in Myriad was a product claim rather than a method claim.[154] The invention, in the opinion of the Full Court, involved a method for distinguishing between maternal and foetal nucleic acids and its practical application.[155] Although the foetal nucleic acid could be seen in nature, cffDNA was not the actual material of the invention. Instead, it was the identification of that specific nucleic acid as an element of a method,[156] which required an artificially created state of affairs and resulted in an economically useful outcome.[157] In light of this, the Full Court concluded that the claims in issue were directed towards patentable subject matter.

Considering the conclusions of Australian courts and the APO in various cases regarding patentable subject matter requirements, methods utilising naturally existing products would be considered patentable as a manner of manufacture. However, for claims to be patentable subject matter, they must satisfy key principles, including involving an artificially created state of affairs and demonstrating economic significance. Additionally, it's important to note that the substance of the claim as a whole must not be naturally existing and should involve human action.

3.3.2.3 Impact on Methods of Developing or Delivering Bioinks or Bioprinted Tissues

Patentable subject matter requirements are applied to patents based on how the subject matter has been claimed, and the outcome of subject matter assessments differs on a case-by-case basis. The analysis conducted in this chapter leads to a preliminary conclusion that, in general, method claims directed towards bioinks and bioprinted tissues, following the principles set by *NRDC*, are likely to be considered a manner of manufacture and may be eligible for patent protection. Furthermore, these method claims would hold economic significance due to their applications in the fields of health and medicine.[158] Similarly, bioinks and bioprinted tissues provide economic advantage in a way that their further application would be of value in a field of 'economic endeavour.'

Considering the analysis and outcomes of the APO's decision in *International Stem Cell Corporation*, it is observed that inventions related to stem cells may not be considered patentable under s 18(2) of the *Patents Act* if they lead to the development of a human being. Living cells and specifically autologous stem cells are the major

---

[153] *Ariosa Diagnostics Inc v Sequenom Inc* [2021] FCAFC 101 at 114.

[154] Ibid., 129.

[155] Ibid., 155.

[156] Ibid.

[157] Ibid., 159.

[158] *AU Patent No. AU2014389440A1*, filed on 06 June 2014 ('Published' on 20 October 2016).

components of bioinks and bioprinted tissues.[159] Bioprinting technology at its current state is only capable of developing human body parts, tissues, and organs at the tissue level.[160] Consequently, most subject matter related to bioinks and bioprinted tissues is unlikely to fall within the scope of s 18(2), as these inventions do not lead to the development of human beings.

Similarly, the FCA's decision in *Cargill* has provided some clarity regarding the patentability of methods of using bioinks and bioprinted tissues. Specifically, considering the analysis of *Cargill*, methods of using bioinks and bioprinted tissues can be considered patentable subject matter. Despite the use of naturally occurring elements such as living cells and growth factors, the bioprinting method involves human interaction to create a balanced composition, allowing living cells to survive and proliferate in external conditions within a bioprinted tissue construct. Both the method and the final product of the bioprinting process would likely be considered as an 'artificially created state of affairs.'

One of the interesting arguments made by MLA with reference to claims 7 and 11 of the Branhaven patent is that these claims are not a manner of manufacture in light of the *D'Arcy* decision as they are directed towards 'a method for cloning a cattle' and 'a cloned cattle' that claims a cloned animal with the same genetic code as the animal it is cloned from.[161] However, Justice Beach rejected MLA's arguments by specifying that '…claim [claim 7] is directed to a practical application, which undoubtedly involves human intervention, and which results in an artificially created state of affairs of economic significance.'[162] Here, the practical application was cloning of a bovine subject. Similarly, with regard to claim 11, Justice Beach rejects MLA's argument, reasoning that an artificial object of economic significance is produced for its own sake, not merely as a receptacle for its informational content.[163] This reasoning provides an interesting perspective on bioprinting, suggesting that a bioprinted tissue, structurally and functionally similar to a natural tissue and created using specific bioprinting techniques, may be considered patentable subject matter.

Drawing a parallel to the *Sequenom* case, where the method of detecting naturally occurring cffDNA was deemed a 'manner of manufacture,' methods of developing

---

[159] Qi Gu et al., 'Stem Cell Bioprinting: Functional 3D Neural Mini-Tissues from Printed Gel-Based Bioink and Human Neural Stem Cells' (2016) 5(12) *Advanced healthcare materials* 1428–1428.

[160] Wei Long Ng, Chee Kai Chua and Yu-Fang Shen, 'Print Me An Organ! Why We Are Not There Yet' (2019) 97 *Progress in Polymer Science*; Ishita Matai et al., 'Progress in 3D Bioprinting Technology for Tissue/Organ Regenerative Engineering' (2020) 226 *Biomaterials* 119,536–119,536.

[161] *AU Patent No. AU2010202253B2* (n 105) Claim 7: 'A method for cloning a bovine subject with a desired trait, comprising: (a) identifying an occurrence of at least three single nucleotide polymorphisms (SNPs) wherein the at least 3 SNPs are associated with the trait, and the at least three SNPs occur 68 in more than one gene, wherein at least one SNP corresponds to position 300 of one of SEQ ID NOS: 19473 to 21982, or the SNP is about 500,000 or less nucleotides from position 300 of any one of SEQ ID NOS: 19473 to 21982; (b) isolating a progenitor cell from the bovine subject; (c) generating a cloned bovine from the progenitor cell.', Claim 11: 'A bovine subject resulting from the method of claim 7.'

[162] *Meat & Livestock Australia Limited v Cargill, Inc* [2018] FCA 51. 470.

[163] Ibid.

bioinks and bioprinted tissues or organs may also be considered patentable, despite utilising naturally existing products such as living cells and growth factors with 'human action.' For example, a patent application filed by Organovo with IP Australia claims a method for developing an implantable tissue, which involves bioprinting bioink containing muscle cells into a form and fusing it into a cohesive cellular structure, provided that the tissue or organ is implantable in a vertebrate subject and not a vascular tube.[164] Considering Justice Beach's analysis and conclusions, such a claim would be considered a manner of manufacture as the steps involved would require human action.

3.4 Recommendations: Patentability Framework Model—Australia

In developing a robust framework for assessing patentability, it is crucial to critically understand the foundational criteria that determine whether a subject matter can be patented. To be considered patentable, subject matter must meet the manner of manufacture requirements, which include 'economic significance,' 'artificially created state of affairs,' and 'human action.' These criteria were clarified by the Australian courts in *National Research Development Corporation v Commissioner of Patents* (*NRDC*) and *D'Arcy v Myriad Genetics* (*D'Arcy*).[165] A bioprinting invention related to bioinks or bioprinted tissues will need to satisfy the above-described criteria to be considered patentable subject matter according to s 18(1) of the *Patents Act 1990*.[166] Specific subject matter with the substance of the claims directed towards living cells or stem cells may not be considered patentable due to the lack of 'human action' and because it does not lead to an 'artificially created state of affairs.'[167] However, subject matter related to bioinks and bioprinted tissues with claims directed towards a combination of natural and artificial elements such as synthetically produced 'living cells' or 'biomaterial' may meet the 'human action' and 'artificially created state of affairs' criteria.[168]

Recommendation 1 (Australia)

The following strategies are recommended to patent applicants to satisfy the manner of manufacture requirements for inventions with claims directed towards bioprinted tissues or bioinks.

[164] *AU Patent No. AU2012308724A1*, filed on 12 September 2012 ('Published' on 24 April 2014).

[165] See Sects. 3.3 and 3.4 of this chapter; *D'Arcy v Myriad Genetics Inc* [2015] HCA 35; *National Research Development Corporation v Commissioner of Patents* (1959) 102 CLR 252.

[166] See Sect. 3.3.1 of this chapter; *Patents Act*, s 18(2).

[167] See Sect. 3.3.1.2 of this chapter.

[168] See Sect. 3.3.2.1 of this chapter.

i. **Artificially created state of affairs**: A patent application with claims directed towards bioprinted tissues or bioinks should include synthetic elements in combination with natural elements to satisfy the requirement of an artificially created state of affairs.
ii. **Human action**: A patent application with claims directed towards bioprinted tissues or bioinks must emphasize the amount of human intellect or intervention that was required to develop something which is not naturally existing. For example, use of human intellect to design a bioink composition and identification of specific elements of bioinks that are essential for better printability and cell viability.
iii. **Economic significance**: A patent application with claims directed towards bioprinted tissues or bioinks must refrain from claiming subject matter that is economically non-significant. For example, a product or composition claim directed towards a bioink may be considered economically significant. However, a method claim directed towards direct bioprinting of such bioink on a patient as a part of a treatment may be considered economically non-significant.

Subject matter with claims directed towards developing bioinks and bioprinted tissues for 'medical treatment' treads a fine line between patentable and non-patentable subject matter when the 'economic significance' criterion is taken into consideration (in light of *NRDC* and *Apotex Pty Ltd v Sanofi-Aventis Australia Pty Ltd* (*Apotex*)).[169] Methods involving doctors (and other medical personnel) physically treating patients are considered to be 'essentially non-economic' and may not be subject to patent protection, according to the conclusions reached in *Apotex*.[170] Subject matter with claims directed towards a method of treating a disease in a human body by implanting a bioprinted tissue or injecting the bioink, which can only be undertaken by physicians, may be considered to be a surgical procedure involving doctors and may not therefore be considered to have 'economic significance.'[171] To build on these suggestions, a patentability framework model has been developed and is visualised in Fig. 3.1. This proposed model is presented to identify the patentability of different types of patent claims relating to the subject matter at issue in Australia.

To elaborate, the applicability of the patentable subject matter requirements under Australian law varies on a case-to-case basis, dependent on how specific subject matter related to bioprinted tissues and bioinks is claimed, highlighting the importance of claim drafting to delineate the claims made in a patent application and ensure they do not fall foul of the 'human action,' 'artificially created state of affairs,' and 'human treatment' requirements for patentable subject matter.

---

[169] See Sect. 3.3.2.1 of this chapter; *Apotex Pty Ltd v Sanofi-Aventis Australia Pty Ltd* [2013] HCA 50.

[170] Ibid.

[171] See Sect. 3.3.2.3 of this chapter.

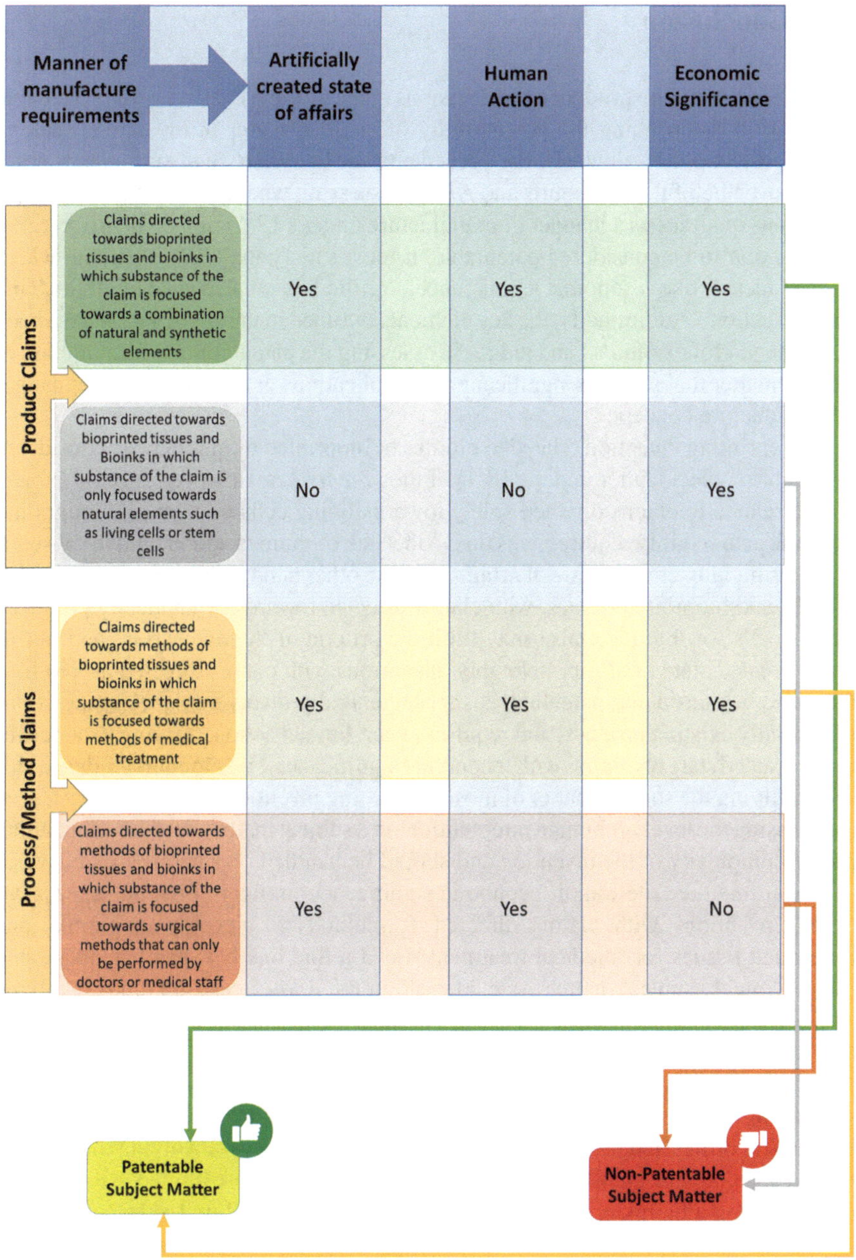

Fig. 3.1 Visualisation of a proposed patentability framework model for patenting inventions related to bioprinted tissues and bioinks in Australia

3.5 Conclusion

Australian patent jurisprudence and decisions made by the APO have offered valuable guidance in determining the patentability of subject matter in inventions seeking patent protection. Particularly, the precedents set by *NRDC* and subsequent cases have proven helpful to the courts and APO in assessing whether the subject matter of inventions qualifies as a manner of manufacture under s 18(1) of the *Patents Act*. For an invention to be considered patentable, it must satisfy the *NRDC* criteria, which include factors like 'economic significance,' 'artificially created state of affairs,' and 'human action.' Additionally, the key elements outlined in the *D'Arcy* case have been instrumental for examiners and judges in assessing the patentability of a new class of subject matter that involves significant new applications or extensions of the manner of manufacture concept.

A bioprinting invention related to bioinks or bioprinted tissues may be considered patentable subject matter under s 18(1) of the *Patents Act*. However, specific subject matter related to claims directed solely towards living cells or stem cells might not meet the patentability requirements due to the lack of 'human action' and the absence of an 'artificially created state of affairs.' On the other hand, subject matter related to bioinks and bioprinted tissues, with claims directed towards synthetically produced 'living cells' or 'biomaterials,' may fulfil the criteria of 'human action' and 'artificially created state of affairs.' Notably, inventions with claims directed at 'bioinks' may also be considered patentable, as these generally involve unique compositions of naturally existing products and require 'expert human action,' leading to an 'artificially created state of affairs' with 'economic significance.' While court guidance aids in identifying the subject matter of most bioprinting inventions, certain aspects, such as assessing the level of 'human intervention' in a subject matter, may vary depending on the complexity of the invention and should be handled on a case-by-case basis.

Regarding the criterion of 'economic significance' in light of *NRDC* and *Apotex* cases, inventions with claims directed at methods of developing bioinks and bioprinted tissues for 'medical treatment' tread a fine line between patentable and non-patentable subject matter. As evidenced in the *Apotex* case, methods involving physical treatment of patients by doctors or medical staff are considered 'essentially non-economic' and may not be eligible for patent protection. For instance, an invention with claims directed at 'a method of treating a disease in a human body by implanting a bioprinted tissue or injecting the bioink,' which can only be performed by medical professionals, may be viewed as a surgical process involving doctors and may lack 'economic significance.' Conversely, if the claims are limited to the method of developing a bioink or bioprinted tissue, they may not fall under the 'procedures of doctors' exclusion and could potentially meet the 'economic significance' criterion. Similarly, considering the conclusions of the *Cargill* and *Ariosa* cases, a claim directed at a method of developing a bioink or bioprinted tissue comprising living stem cells may be seen as a practical application of a naturally occurring element or phenomenon (stem cells) and could be considered patentable if the claimed invention results in an 'artificially created state of affairs' with an economically useful outcome.

Although it is obvious to claim that there is significant human action and effort in the development of bioinks and bioprinted tissues, the *NRDC* principles did not shed light on how significant 'human action' must be to consider an invention to be an 'artificially created state of affairs.' Although the *D'Arcy* decision has provided some useful guidance, there is no one straightforward test to determine whether or not a claimed invention is a manner of manufacture. Specifically, with reference to method claims, Australian courts have provided different opinions that lead to some uncertainty in identifying the scope of patentable subject matter requirements. The empirical component of this book considers the gaps identified in this chapter (such as the ambiguity and uncertainty in the application of current provisions) and offers a patent prosecution analysis of individual patent applications with claims relating to bioprinted tissues and bioinks in order to shed light on how the current provisions are being applied by patent offices when evaluating particular bioprinting inventions.

Chapter 4
Are 3D Bioprinted Tissues and Bioinks Patentable Subject Matter in the US?

4.1 Introduction

The patent system enshrined in the US Constitution was a far-sighted idea and allowed everyone to contribute to the country's development. The US Constitution (1787) (Article 1, Section 8) contains the following paragraph: 'to promote the progress of science and useful arts, by securing for limited times to authors and inventors the exclusive right to their respective writings and discoveries.'[1] The US intellectual property industry provides 40 million jobs and contributes $5 trillion to gross domestic product (almost 35%).[2] The importance of intellectual property for the economy has become a motivating factor for the USPTO to take the initiative to improve the quality of patents.[3] Similarly, US patent law has been adapting to accommodate advanced technologies to protect and promote innovation and at the same time enable incentivisation for the intellectual efforts of inventors.[4] The *Leahy-Smith America Invents Act 2011* (AIA) is the most significant development in US

[1] *The Constitution of the United States*, created on 17 September 1787, ratified on 21 June, 1788.

[2] USPTO Report Shows Intellectual Property-Intensive Industries Contribute $5 Trillion, 40 Million Jobs to U.S. Economy, *USPTO*, https://www.uspto.gov/ip-policy/economic-research/publications/reports/uspto-report-shows-intellectual-property-intensive.

[3] Patent quality, *USPTO*, https://www.uspto.gov/patents/patent-quality. '*Providing high-quality, efficient examination of patent applications is paramount to our mission at USPTO. To ensure we continue to issue high-quality patents that will fuel innovation well into the future, the Office of the Deputy Commissioner for Patent Quality, along with our partners across the Patents organization, promotes and supports the continuous improvement of patent products, processes and services through collaboration with internal and external stakeholders of the intellectual property community*'.

[4] R. Levin, 'A Patent System for the 21st Century' (2004) 20(4) *Issues in Science and Technology* 49–54.

P. Devarapalli, *Bioinked Boundaries*, https://doi.org/10.1007/978-981-96-2591-8_4

patent law in the last 60 years (since the adoption of the AIA in 1952).[5] The further sections of this chapter provide an extensive analysis of the patentable subject matter requirement in the US, specifically in light of subject matter related to bioprinted tissues and bioinks. This chapter will examine the relevant statutory provisions, judicial interpretations, and guidelines that shape the scope and boundaries of patent eligibility for biotechnological inventions from a bioprinting perspective.

4.2 Overview of Patentable Subject Matter Requirements in US

The conception of patentable subject matter has remained largely unchanged under US law for the past two centuries.[6] The US Congress has traditionally taken an all-inclusive approach to define patentable subject matter, reflecting the US Constitution. This all-inclusive approach is evident in the *Patent Act* 1793 (the first patent law) of Congress, which described patentable subject matter as 'any new and useful art, machine, production [,] or composition of matter, or any new and useful enhancement[thereof].'[7] This wording was not changed until Congress replaced the word 'art' with 'process' in the 1952 *Patent Act*. However, technology has changed dramatically during this period, evolving from purely mechanical inventions of the industrial revolution into a multidimensional landscape. Most inventions in the age of the industrial revolution were mechanical. This has gradually changed due to the advancement of science and technology.[8] For example, technological advancements in computers, information technology, biotechnology, and nanotechnology have raised questions about the adaptability of existing patent laws to new innovations.[9] While an all-inclusive approach provides flexibility, overly prescriptive language may hinder adaptation to emerging technologies.

In the US, the subject matter of an invention, regardless of being new and useful, should also be within the boundaries of patentable subject matter set forth by Section 101. While the patent eligibility requirement under Section 101 is highly crucial (particularly for subject matter related to biotechnology, software, and business methods), there is very little guidance in the actual statutory language:

---

[5] Paul R. Gupta and Alex Feerst, 'The U.S. Patent System After the Leahy-Smith America Invents' reprinted from E.I.P.R. Issue 1, 2012, URL https://www.orrick.com/Events-and-Publications/Documents/4466.pdf.

[6] Robert Greene Sterne and Lawrence B Bugaisky, 'The Expansion of Statutory Subject Matter Under the 1952 Patent Act' (2004) 37 *Akron Law Review* 217.

[7] U.S. Constitution, Article. I, § 8, cl. 8.

[8] Christopher Freeman et al., *As time goes by: from the industrial revolutions to the information revolution* (Oxford University Press, 2001).

[9] John H. Barton, 'Adapting the intellectual property system to new technologies' (1995) 10(2–3) *International Journal of Technology Management* 151–172.

> Whoever invents or discovers any new and useful process, machine, manufacture, or composition of matter, or any new and useful improvement thereof, may obtain a patent therefor, subject to the conditions and requirements of this title.[10]

It can be noted that Section 101 provides four classes of patentable subject matter, that is (i) process; (ii) machine; (iii) products of manufacture; and (iv) composition of matter.[11] There is some similarity in relation to the use of the word 'manufacture' between the US and Australian provisions on patentable subject matter requirements. Specifically, as discussed in Chap. 3, s 18(1) of the Australian *Patents Act 1990* highlights the requirement that subject matter must meet the manner of manufacture requirements to be considered patentable. While the provisions of Section 101 appear more structured by providing specific types of subject matter, the Australian approach may be considered to have a broader scope considering the definition of manner of manufacture as analysed in Chap. 3.

The Manual of Patent Examining Procedure (MPEP) sets forth two conditions for assessing subject matter eligibility.[12] Firstly, the claimed invention must fall under one of the four legislative categories: process, machine, manufacture, or composition of subject matter. Secondly, the claimed invention must not fall within a judicial exception, which includes abstract theories, laws of nature, and natural phenomena, including products of nature. Judicial exceptions are subject matter which the courts have considered to be beyond, or exceptions to, the four legislative categories of invention and are restricted to abstract theories, laws of nature, and natural phenomena (including products of nature).[13]

The judicial exception of abstract theories was derived from the historical case *Mackay Radio & Telegraph Co. v. Radio Corp. of America* (*Mackay*). In this case, the court held that 'a scientific truth, or the mathematical expression of it,' is not patentable, but a novel and useful structure created using scientific knowledge may be patented.[14] Similarly, the judicial exception of laws of nature and natural phenomena was identified in 1948 in the case *Funk Bros. Seed Co. v. Kalo Co* (*Funk Bros.*). In this case, the court investigated whether the alleged application of the natural phenomenon was inventive or merely necessary standard skills in the field of art. As there was no invention in the application of the discovered phenomena, the subject matter of the invention was considered unpatentable.[15] Since these decisions, subject matter related to abstract theories, laws of nature, and natural phenomena has been excluded from patentability in the US. Considering these two decisions, different courts have applied the exception in different cases. Some of these cases

---

[10] *Leahy-Smith America Invents Act 2011* (US) s 101; ('The term "process" means process, art, or method and includes a new use of a known process, machine, manufacture, composition of matter, or material.').

[11] Ibid.

[12] United States Patent and Trademarks Office (2019), *Manual of Patent Examining Procedure*, Part 2106.

[13] Ibid.

[14] *Mackay Radio & Telegraph Co. v. Radio Corp. of America*, 306 U.S. 86 [1939).

[15] *Funk Bros. Seed Co. v. Kalo Co.*, 333 US 127 [1948].

will be discussed in detail in the following sections of this chapter. Comparing this to the Australian perspective as discussed in Chap. 3, although the *Patents Act 1990* provides a number of specific exclusions from patentability, those exclusions are limited and can be interpreted narrowly in comparison with the exclusions provided under US law.[16]

On its face, subject matter directed towards bioinks and bioprinted tissues as products may be considered as patentable subject matter as they may be classified as products of manufacture. Similarly, methods or processes related to bioinks and bioprinted tissues may also be considered as patentable subject matter as they may be classified as processes of manufacture. However, the presence of natural elements in claims may invoke the products of nature doctrine and exclude specific subject matter related to bioinks and bioprinted tissues.[17] Based on the analysis presented in Chap. 3 of this book, applying the US concept of products of nature to inventions concerning bioprinted tissues and bioinks could yield a similar outcome. If an invention with claims directed at bioprinted tissues and bioinks meets the requirements of manner of manufacture, such as involving human action and resulting in an artificially created state of affairs, it may not be regarded as a natural phenomenon or product of nature according to the US perspective. This is due to the crucial element of human action leading to the creation of an artificially produced state. The subsequent sections of this chapter will focus on evaluating whether subject matter that claims bioinks and bioprinted tissues as products and processes are likely to be considered patentable subject matter, taking into account relevant US provisions and jurisprudence.

4.3 Analysis of US Patent Jurisprudence Discussing Patentable Subject Matter

Over the years, US patent jurisprudence has provided some guidance on patentability of subject matter related to advanced technologies such as biotechnology and information technology.[18] In the US, patent applications are initially assessed by patent examiners at the USPTO and if specific patent claims are not considered patentable by the patent examiner, the applicant is allowed to appeal to the Patent Trial and Appellate Board (PTAB) by filing a notice of appeal and an appeal brief. Similarly, patent applications that are approved by the USPTO can also be challenged at the PTAB in the form of post-grant review. Where patent claims are rejected by the USPTO and again rejected at the PTAB, the applicant may appeal the decision of PTAB at the District Court and Court of Appeals of the Federal Circuit levels, followed by a final appeal to the US Supreme Court. Decisions of the US Supreme Court have the

[16] See Sect. 3.4.2 of Chap. 3.

[17] N. Ashammakhi et al., 'Bioinks and bioprinting technologies to make heterogeneous and biomimetic tissue constructs' (2019) 1 *Materials Today Bio*.

[18] Michael S. Mireles, 'Patent-Eligible Subject Matter in the United States: An Evolving Landscape', *Intellectual Property Issues in Microbiology* (Springer, 2019) 3–35.

highest precedential value as this Court is the top in the hierarchy of the US judicial system.[19] This section of the chapter conducts an analysis of pertinent US case law, with a specific focus on decisions made by the Supreme Court. It examines how these precedents might impact the patent eligibility of subject matter related to products of bioprinting, such as bioinks and bioprinted tissues.

As mentioned in the previous section, judicial scrutiny of patentable subject matter dates back to 1939 with the *Mackay* case, followed by the *Funk Bros.* case in 1948. These two decisions offered guidance on which types of inventions should not be deemed patentable subject matter. However, the questions that arise from this are how to determine the threshold for patentability and what additional factors are required to establish subject matter as patentable. In the *Gottschalk v. Benson* case, the US Supreme Court examined whether an algorithm for converting binary-coded decimal numbers into real binary numbers could be considered patentable. The Court expressed concern that granting a patent for this algorithm might hinder the use of the concept for developing new inventions. As a result, the Supreme Court ruled that this algorithm was not patentable, as abstract theories and mathematics should be treated as abstract concepts and not eligible for patent protection.[20]

A similar analysis was employed in the case of *Parker v. Flook* in 1978, where the US Supreme Court examined the patentability of a system used to update alarm limits on catalytic converters. These alarm limits are specific values that define the normal operating range of a catalytic converter, an exhaust emission control device used in motor vehicles. An alarm is triggered if the values deviate from this range. The patent claims in question focused on these alarm limits themselves, and the only distinction between the prior art and the invention was the algorithm used to calculate the new alarm limit.[21] The Court found that the claims directed at the algorithm were not patentable, even though they included an additional step *beyond* the measurement process. The Court reiterated that algorithms are mathematical concepts or mental processes that do not go beyond abstract ideas. In light of this, the Court explicitly rejected the notion that 'post-solution activity[22] [alone]… can transform an unpatentable principle into a patentable process.' The Court clarified that the rejection was not solely because the claim contained a mathematical algorithm as one component, but rather because the application, taken as a whole, lacked any patentable invention once that algorithm was assumed to be part of the prior art.

This decision made it quite clear that inventions claiming a mathematical algorithm are not regarded as patentable subject matter because it is assumed that the

[19] Ali Masood, 'The Impact of Supreme Court Precedent in a Judicial Hierarchy' (PhD Thesis, University of South Carolina, 2016) https://scholarcommons.sc.edu/etd/3906/.

[20] *Gottschalk V. Benson*, 409 U.S. 63 [1972].

[21] *Parker v. Flook*, 437 U.S. 584 [1978].

[22] An example of a pre-solution activity is a step that is recited as part of a claimed process of analysing and manipulating the acquired information by a series of steps in order to determine if the transactions were fraudulent. This step might involve obtaining information regarding credit card transactions.

claimed algorithm is within the prior art. In addition, the unpatentability of algorithms at USPTO is primarily due to the fact that algorithms include a series of mathematical steps and procedures and thus are considered abstract.

4.3.1 Recombinant Bacteria—Chakrabarty

A few years later, in 1980, the US Supreme Court analysed the patentable subject matter provisions (Section 101 of the AIA) in the *Diamond v. Chakrabarty* case (*Chakrabarty*).[23] Anand Chakrabarty, a microbiologist, filed a patent application and assigned it to the General Electric Co. The patent application comprised 36 claims, encompassing a recombinant oil-degrading bacterium, a method of producing the bacteria, and an inoculum with the new recombinant bacteria on a carrier material floating on water, such as straw. While the claims directed towards the process and inoculum were allowed, the claims concerning the new recombinant bacteria faced rejection. The examiner's reasoning was two-fold: first, micro-organisms were deemed 'products of nature,' and second, living things, including micro-organisms, were considered non-patentable subject matter under Section 101. Chakrabarty appealed the denial of these claims to the Board of Appeals of the Patent Office, but the Board upheld the examiner's decision, stating that legislative history indicated Section 101 did not encompass living things, even if created in laboratories, based on the *Plant Patent Act* of 1930 providing patent rights to asexually reproduced plants.

The decision of the Board was further appealed to the Court of Customs and Patent Appeals (CCPA), which examined the subject matter in light of the previous decision in *Parker v. Bergy* (*Bergy I*) that overturned the judgement of the Board and extended patent rights to living organisms.[24] The same appellate court reversed the Board's decision in *Chakrabarty* on March 2, 1978, allowing the grant of patent rights for living organisms.[25] The appellate court based its judgement solely on its earlier ruling in *Bergy I*.[26] Although a petition for writ of certiorari for the *Bergy I* case was issued by the Supreme Court, the case was eventually returned to the CCPA for re-examination in the light of the ruling of the Supreme Court in *Parker v. Flook*.[27] The Commissioner for Patents and Trademarks then petitioned the CCPA to set aside its ruling on *Chakrabarty* based on *Parker v. Flook* and the order of the

[23] *Diamond v. Chakrabarty*, 447 U.S. 303, 309 [1980].

[24] *Parker v. Bergy*, 563 F.2d 1031 (C.C.P.A. 1977) [*Bergy I*]. In *Bergy*, the court found a living organism to be patentable subject matter according to Section 101. The microbiological process in *Bergy I* was based on '*Streptomyces vellosus*', a bacterial strain that could produce an antibiotic called Lincomycin in a recoverable quantity through a fermentation process.

[25] The Board held that Chakrabarty's claims were invalid and not statutory subject matter according to Section 101.

[26] *Bergy*, 571 F.2d 40 (C.C.P.A. 1978). This was the appellate court opinion for *Diamond v. Chakrabarty*, 100 S. Ct. 2204, which is the subject of this case comment.

[27] *Parker v. Flook*, 437 U.S. 584 [1978].

Supreme Court of *Bergy I*.[28] The CCPA combined the judgements of *Bergy I* and *Chakrabarty* and confirmed the earlier two decisions by specifying that *Parker v. Flook* had no impact on either case. Although certiorari was again granted in both *Bergy I* and *Chakrabarty*, only *Chakrabarty* was taken up for review by the Supreme Court, and *Bergy I* had become moot with no reasoning from the Supreme Court.[29] This rather convoluted process led to what is perhaps one of the most important decisions on products of nature in the US.

The Supreme Court saw the question as a narrow one of statutory interpretation, focusing on construing Section 101. It aimed to determine whether Chakrabarty's recombinant micro-organism fell within the categories of 'manufacture' or 'composition of matter' as defined in the statute.[30] Chief Justice Burger led the analysis and authored the majority decision. The Court defined 'manufacture' as the production of articles from raw materials with new forms, qualities, properties, or combinations, whether by hand, labour, or machinery.[31] Similarly, 'composition of matter' included all compositions of two or more substances, whether resulting from chemical union or mechanical mixture, encompassing gases, fluids, powders, or solids.[32] The Chief Justice emphasised that Congress intended for patent law to have a broad scope, covering a variety of subject matter for patent protection.[33]

The Chief Justice supported the broad construction of 'manufacture' or 'composition of matter' by referring to the relevant legislative history, which showed its evolution and how Congress maintained the wording to provide a broad scope to patent laws.[34] Having construed the language with this broad scope, the Court also noted the limitations imposed by judicial exceptions saying, 'This is not to suggest that § 101 has no limits or that it embraces every discovery. The laws of nature, physical phenomena, and abstract ideas have been held not patentable.'[35]

---

[28] Ibid.

[29] *Diamond v. Chakrabarty*, 444 U.S. 1028 [1980].

[30] Ibid.

[31] *Diamond v. Chakrabarty*, 444 U.S. 1028 [1980], III; *American Fruit Growers, Inc. v. Brogdex Co.*, 283 U. S. 1, 11 [1931]. (a case involving a claim for a new process, which used a chemical solution for preparing fresh fruit for market); According to the Oxford Dictionary the definition of the word 'manufacture' is 'An article, material, or commodity produced by physical labour, machinery, etc.'

[32] *Shell Development Co. v. Watson*, 149 F. Supp. 279, 280 [DC 1957] (citing 1 A. Deller, Walker on Patents § 14, p. 55 (1st ed. 1937)).

[33] However, it is also important to note that the inventions that are identified as a manufacture' or 'composition of matter' should also have a subject matter which is novel, non-obvious, and useful.

[34] The Patent Act of 1793, authored by Thomas Jefferson, defined statutory subject matter as 'any new and useful art, machine, manufacture, or composition of matter, or any new or useful improvement [thereof].' Act of Feb. 21, 1793, s 1, 1 Stat. 319. Subsequent patent statutes in 1836, 1870, and 1874 employed this same broad language. In 1952, when the patent laws were recodified, Congress replaced the word 'art' with 'process,' but otherwise left Jefferson's language intact. The Committee Reports accompanying the 1952 Act inform us that Congress intended statutory subject matter to 'include anything under the sun that is made by man.' S. Rep. No. 1979, 82d Cong., 2d Sess., 5 (1952); H. R. Rep. No. 1923, 82d Cong., 2d Sess., 6 (1952).

[35] *Diamond v. Chakrabarty*, 444 U.S. 1028 [1980], III.

He held that the micro-organism itself was a patentable subject matter under Section 101. The Court stated that '…the patentee has produced a new bacterium with markedly different characteristics from any found in nature and one having the potential for significant utility. His discovery is not nature's handiwork, but his own….' This approach is akin to the Australian 'artificially created state of affairs' and 'human action' criteria as per NRDC (see Chap. 3). Both jurisdictions allow patenting natural products or processes that are modified by human ingenuity.

The Court ruled that the recombinant bacterium in the invention was a non-natural creation with different characteristics and potential utility than any natural bacteria. The Chief Justice stressed that the invention involved clear human intervention and was not a creation of nature. Additionally, the Chief Justice held that the claimed subject matter directed towards the micro-organism as claimed was a non-naturally occurring manufacture or composition of matter, a product of human ingenuity, and was covered within the language of Section 101 and hence the claim directed towards micro-organism itself qualified as patentable subject matter.[36] The Court concluded that, '…the patentee has produced a new bacterium with markedly different characteristics from any found in nature and one having the potential for significant utility. His discovery is not nature's handiwork, but his own….'[37] This approach is similar to the Australian approach of 'artificially created state of affairs' and 'human action' according to *National Research Development Corporation* (*NRDC*) as detailed in Chap. 3.[38] Both jurisdictions consider subject matter related to nature or products of nature patentable when the claimed process or product is an output of human intervention.

The Chakrabarty ruling exemplifies Congress's intent for Section 101 to encompass diverse subject matter involving human ingenuity.[39] The decision has not only had a huge impact on biotechnological inventions, but it has also opened doors for any new subject matter having human ingenuity, by expanding the scope of patentable subject matter and changing the way Section 101 was interpreted.[40] For example, an interpretation of the phrase 'any new' in Section 101 in light of the Chakrabarty decision would enable the patentability of claims directed towards natural elements with markedly different characteristics as a result of human ingenuity.

The Chakrabarty decision holds significant relevance to subject matter related to bioprinted tissues and bioinks. Applying the *Chakrabarty* framework, claims directed towards bioinks and bioprinted tissues, if they involve significant human intervention and result in novel, non-naturally occurring compositions or processes, could potentially be considered patentable under Section 101 in the US.

---

[36] Ibid.

[37] Ibid.

[38] See Sect. 3.4.2 of Chap. 3.

[39] Lumelsky, Anna, 'Diamond v. Chakrabarty: Gauging Congress's Response to Dynamic Statutory Interpretation by the Supreme Court' (2004) 39 *University of San Francisco Law Review* 641.

[40] Darr, Frank P., 'Expanding patent coverage: policy implications of Diamond v. Chakrabarty' (1981) 42 *Ohio State Law Journal* 1061.

4.3.2 Computer Software—Diehr

A year after the *Chakrabarty* decision, in 1981, the Supreme Court delivered another decision on the interpretation of Section 101 in the case of *Diamond v. Diehr* (*Diehr*).[41] The subject matter in this case involved a system for deciding how rubber can be heated in order to be properly 'cured.' The invention used a device to measure and monitor the heating periods for the rubber. The invention (as specified in the claims) comprised not only the computer software but also the steps related to the heating of rubber and the extraction of rubber from heat. The Supreme Court ruled that, in this case, the subject matter was not merely a mathematical algorithm, but a rubber moulding process, and was therefore patentable. The *Diehr* case is widely considered as one of the landmark decisions for subject matter related to computer and software-related inventions.[42] While the decision in *Diehr* was specifically related to computer and software-related inventions, the principles established by *Diehr* have broader implications. For example, if a specific claim recites a mathematical formula, scientific principle, or natural phenomenon and employs the same to transform or reduce an article to a different state or thing, the claim satisfies the requirements of Section 101 as provided by *Diehr*. This has broadened the scope of patentable subject matter, and it has made it easier to patent inventions that involve computer software.

Although the *Diehr* decision may not be directly relevant in the present context of bioinks and bioprinted tissues, together, the *Chakrabarty* and *Diehr* Supreme Court decisions, in consecutive years, brought great clarity to interpretation and application of Section 101 in relation to a variety of subject matter.[43] A few years after the *Chakrabarty* and *Diehr* decisions, on 07 April 1988, a statement was issued by USPTO announcing that, 'the Patent and Trademark Office now considers non-naturally occurring non-human multicellular living organisms, including animals, to be patentable subject matter within the scope of Section 101.'[44] Almost a year later, on 12 April, 1988, the USPTO issued the first patent for a transgenic animal, also known as the 'Harvard Mouse.'[45] One of the major reasons behind the USPTO considering non-naturally occurring non-human multicellular living organisms, including animals, to be patentable subject matter was based on their markedly different characteristics, as concluded in the *Chakrabarty* case.

---

[41] *Diamond v. Diehr*, 450 U. S. 175 [1981].

[42] Nimtz, R. O., 'Diamond v. Diehr: a turning point' (1980) 8 *Rutgers Computer & Technology Law Journal* 267.

[43] Robinson, Douglas and Nina Medlock, 'Diamond v. Chakrabarty: a retrospective on 25 years of biotech patents' (2005) 17(10) *Intellectual Property & Technology Law Journal* 12–15; Stout, Donald E., 'Protection of Programming in the Aftermath of Diamond v. Diehr' (1983) 4 *Computer/LJ* 207.

[44] US Patent and Trademark Office Notice, 'Animals–Patentability,' 1077 O.G. 24 (Apr. 21, 1987).

[45] *US Patent No. US4736866A*, filed on 22 June 1984 ('Granted' on 12 April 1988).

4.3.3 Method of Using Excluded Subject Matter—Mayo

In the 2012 *Mayo Collaborative Services v. Prometheus Laboratories, Inc.* (*Mayo*) decision, the US Supreme Court addressed the patentability of a process. The Court ruled that the claimed subject matter involved laws of nature, which are not patentable.[46] Laws of nature and natural phenomena encompass principles, relations, or products derived from nature, whether naturally occurring or lacking markedly different characteristics from natural occurrences.[47] The case focused on patent claims related to procedures aiding doctors in treating autoimmune diseases with thiopurine drugs by determining optimal dosage levels.[48] These claims were found to apply natural laws concerning the correlation between specific thiopurine metabolite concentrations in the blood and treatment effectiveness or adverse effects.[49] Past precedents established that overly restrictive process claims or claims solely directed to laws of nature are ineligible for patents.[50] However, the Court acknowledged that claims incorporating laws of nature could be patentable if they involve an 'inventive concept,' beyond the natural law, to ensure the patent amounts to more than just claiming the law of nature alone.[51]

The Court's conclusion that the claims in *Mayo* were limited to natural laws rested on two factors. Firstly, the claims must not preempt the complete implementation of the natural law. Secondly, any additional elements beyond the natural law must be relevant and not mere routine, conventional steps.[52] The Court questioned whether the patent claims added enough to the natural law correlations to qualify as patent-eligible processes that applied natural laws. In this case, the Court's answer was no.[53]

Considering this, and the conclusions of *Diehr* and *Flook*, the Court decided that the process claims in *Mayo* did not satisfy the requisite conditions and, in particular, the steps claimed in the process (apart from the laws of nature alone) involved well-understood, routine, conventional activity previously engaged in by researchers in the field. This decision demonstrates that, in order for a claim to satisfy Section 101, the remaining elements must be sufficiently broad so as to not apply a law of nature that has already been identified. Particularly, they cannot be well-understood, routine, conventional activities already engaged in by the scientific community or, when seen as a whole, activities that add nothing significant in addition to the sum of their parts taken separately. However, the judgement makes no reference to how the steps involved in a process might be utilised to limit the application of a natural law in a

---

[46] *Mayo Collaborative Services v. Prometheus Laboratories, Inc.*, 132 S.CT. 1289 [2012].

[47] United States Patent and Trademarks Office (2019), *Manual of Patent Examining Procedure*, Part 2106.04(b).

[48] *Mayo Collaborative Services v. Prometheus Laboratories, Inc.*, 132 S.CT. 1289 [2012].

[49] Ibid.

[50] *Gottschalk v. Benson.,* [S. CT. 1972].

[51] *Parker v. Flook* (S.CT. 1978) at III; *Bilski v. Kappos* 561 U.S. 593 (2010).

[52] *Mayo Collaborative Services v. Prometheus Laboratories, Inc.*, 132 S.CT. 1289 [2012], II A.

[53] Ibid., II.

way that qualifies a claim for patent protection. Following the *Mayo* decision, the USPTO published updated examination guidelines on patentable subject matter that are further elaborated in Sect. 4.4 of this chapter.

Considering the precedents set by Diamond v. Diehr and Parker v. Flook, the Court found that the process claims in Mayo did not meet the necessary conditions. The steps claimed in the process, apart from the laws of nature, involved well-understood, routine, and conventional activities commonly performed by researchers in the field. The Mayo decision highlights that for a claim to satisfy Section 101, it must encompass elements beyond identified natural laws and cannot solely rely on routine activities already well-known in the scientific community. However, the decision did not provide sufficient guidance on how the steps involved in a process could be used to limit the application of a natural law in a way that qualifies the claim for patent protection. Following Mayo, the US Patent and Trademark Office (USPTO) issued updated examination guidelines on patentable subject matter, further elaborated in Sect. 4.4 of this chapter.

This decision has implications for subject matter related to bioprinted tissues and bioinks, which specifically claims products derived from nature that may involve natural laws or phenomena. For example, bioinks may contain natural biomolecules or cells that have specific functions or interactions. Bioprinted tissues may rely on natural processes such as tissue regeneration or differentiation. To be patentable, these products must have markedly different characteristics from what occurs in nature and must not claim the natural laws or phenomena alone. Additionally, any process claims related to bioprinting or bioink preparation must have an 'inventive concept' that adds something significant to the natural law or phenomenon.

4.3.4 *Genetic Material—Association of Molecular Pathology (AMP)*

Patenting of genetic material and gene sequences has always been contentious around the world.[54] Analysis of decisions related to biological material such as gene sequences and related subject matter is highly relevant in the context analysis patenting aspects of bioinks and bioprinted tissues as they are interrelated and connected to a common point which is 'products of nature.' Hence, further analysis in this chapter will primarily focus on the case decisions related to biological material (and its use) delivered after *Chakrabarty*.

Following the decision of *Chakrabarty,* the USPTO provided a series of guidelines for patenting biological material including genes, genetic material, and DNA

[54] Rebecca S. Eisenberg, 'Why the gene patenting controversy persists' (2002) 77(12) *Academic Medicine* 1381–1387; Timothy Caulfield et al., 'Evidence and anecdotes: an analysis of human gene patenting controversies' (2006) 24(9) *Nature Biotechnology* 1091–1094.

sequences.[55] In the 1990s, there was ongoing debate about the role of patent law in promoting progress in the field of genetics. The level of usefulness or utility that had to be proven to justify an argument over a genetic innovation was one of the points of contention. In 1991, the National Institutes of Health (NIH) filed patent applications for about 2,700 Express Sequence Tags (ESTs). ESTs are a short sub-sequence of a cDNA sequence. ESTs may be used to identify gene transcripts and are instrumental in gene discovery and in gene-sequence determination.[56] The applications included claims directed towards the ESTs and their gene sequences and derivative proteins in their entirety. These applications were controversial because at the time of filing, the functions of these sequences were unknown.[57]

In 1997, the biological materials debate intensified when researchers at Miami Children's Hospital were granted a patent for a gene sequence associated with Canavan's disease. This patent allowed them to restrict other laboratories from conducting a genetic test related to the Canavan gene.[58] Such gene patents, including those for the BRCA genes linked to breast cancer risk, raised concerns about their impact on public health.[59] Myriad Genetics received the first patent for one of the BRCA genes three years later, leading to further controversy over gene patenting, as it directly claimed genes that are commonly considered products of nature.

AMP is an American non-profit professional society focused on the field of molecular diagnostics. In 2010, AMP brought a suit in the US District Court against Myriad's patents having claims directed towards the BRCA1 and BRCA2 genes.[60] Claim 1 of one of the patent applications in issue reads as,

> An isolated DNA comprising an altered BRCA1 DNA having at least one of the alterations set forth in Tables 12A, 14, 18 or 19 with the proviso that the alteration is not a deletion of four nucleotides corresponding to base numbers 4184-4187 in SEQ. ID. NO:1.[61]

Considering the public interest of this litigation, researchers from Columbia, New York University, Emory, and Yale, as well as a number of patient advocacy organisations and individual patients, were among the other plaintiffs in the lawsuit along with the AMP and the University of Pennsylvania.

In the District Court, Sweet DJ conducted a thorough examination of relevant case law concerning subject matter related to purified natural products. He referenced the

[55] Jacob S. Sherkow and Henry T. Greely, 'The history of patenting genetic material' (2015) 49 *Annual Review of Genetics* 161–182.

[56] Parkinson, John and Mark Blaxter, 'Expressed sequence tags: an overview' (2009) *Expressed sequence tags (ESTs)* 1–12.

[57] Molly A. Holman and Stephen R. Munzer, 'Intellectual property rights in genes and gene fragments: A registration solution for expressed sequence tags' (1999) 85 *Iowa Law Review* 735.

[58] Debra L. Greenfield, 'Greenberg v. Miami Children's Hospital: Unjust enrichment and the patenting of human genetic material' (2006) 15 *Annals Health L.* 213.

[59] Patricia A. Lacy, 'Gene patenting: universal heritage vs. reward for human effort' (1998) 77 *Or. L. Rev.* 783.

[60] At issue are claims 1, 2, 5, 6, and 7 of U.S. Patent 5,747,282 (the '282 patent), claim 1 of U.S. Patent 5,693,473 (the '473 patent), and claims 1, 6, and 7 of U.S. Patent 5,837,492 (the '492 patent); *Myriad (District Court)*, 702 F Supp 2d 181, 223–4 [2010].

[61] *US Patent No. US5693473A*, filed on 07 June 1995 ('Granted' on 02 December 1997).

American Wood-Paper Co v Fibre Disintegrating Co (American Wood-Paper) decision, which involved claims directed towards refined cellulose resulting from the decomposition of the original substance. In that case, the Court deemed the claims to encompass non-patentable subject matter since the product existed in nature and was in a nearly pure state. Sweet DJ also cited the *General Electric Co v De Forest Radio Co* which invalidated a patent claiming the element tungsten.[62] Furthermore, the District Court referred to the Supreme Court decision in *American Fruit Growers Inc. v Brogdex Co*, where claims directed towards fruits treated with mould-resistant borax were deemed non-patentable, as the mere addition of borax did not create a product with a new or distinctive form, quality, or property. Based on this analysis, the District Court concluded that a patented manufacture must involve transformation, resulting in a new and different article with a distinct name, character, or use.[63]

In *Merck & Co Inc. v Olin Mathieson Chemical Co (Merck)*, the Federal Court held that vitamin B_{12} (a known compound) was patentable as 'more than a "mere advance in the degree of purity of a known product"'[64] and invigorated Sweet DJ's perspective that the ultimate question was whether 'a new or distinctive form, quality, or property' had emerged.[65] The Court's approach to complementary DNA (cDNA)[66] contradicted the treatment of DNA and RNA as information, rather than molecules. The Court held that 'the fact that the BRCA1/2 cDNA molecules covered by the composition claims-in-suit contain only the protein coding exons and not the introns found in native DNA does not render these cDNAs and their native counterparts "markedly different".'[67] It did not matter that the claimed cDNA does not normally occur naturally because the information does, in the form of mRNA.[68] The nucleotide sequence, if not the molecule itself, is dictated by nature. In light of the above reasoning, the Court therefore considered cDNA unpatentable.

In 2012, the case was appealed, and Lourie J, one of the three judges hearing the case, deemed the isolated DNA claims patent eligible. He held that these claims covered molecules that were markedly different with a distinctive chemical structure

---

[62] *American Wood-Paper Co v Fibre Disintegrating Co*, [1874] USSC 189; 90 US 566, 594 [1874] ('*American Wood-Paper*'); *General Electric Co v De Forest Radio Co*, 28 F 2d 641 [1928] ('*General Electric*').

[63] *American Fruit Growers Inc v Brogdex Co*, 283 US 1, 11 [1931] ('*American Fruit Growers*'): treatment of the outer skin of fruit with mould inhibitor did not convert the treated fruit into an article of manufacture.

[64] *Merck & Co Inc v Olin Mathieson Chemical Co*, [1958] USCA4 73; 253 F 2d 156, 164 [1958].

[65] *American Fruit Growers*, 283 US 1, 11 [1931]; *Myriad (District Court)*, 702 F Supp 2d 181, 227 [2010].

[66] Complementary DNA (cDNA) is DNA that is produced from a single-stranded Ribonucleic Acid (RNA) template using the enzyme reverse transcriptase. In molecular biology, cDNA is also produced to examine the transcriptome patterns in tissue, single cells, or single nuclei in experiments like microarrays and RNA-seq.

[67] Messenger RNA (mRNA) is a single-stranded molecule of RNA that corresponds to the genetic sequence of a gene and is read by a ribosome in the process of synthesising a protein; Complementary DNA (cDNA) is DNA synthesised from a single-stranded RNA; *Myriad (District Court)*, 702 F Supp 2d 181, 231 [2010] at 3.

[68] Ibid., 230.

and identity from those found in nature.[69] Lourie J distinguished this case from *In re Marden et al.* (*Marden*) and *Parke-Davis & Co v HK Mulford Co* (*Parke-Davis*), stating that isolated DNA was not merely purified DNA but a distinct chemical entity obtained through human intervention.[70] The Court further supported this analysis by explaining that the claimed isolated DNA molecules did not exist in nature within a physical mixture to be purified; instead, they required chemical cleaving from their native combination with other genetic materials. Therefore, the Court in *Myriad* viewed DNA as a different molecule obtained through human intervention,[71] focusing on its distinctiveness as 'isolated compositions of matter' rather than its informational content or physiological use.[72] Consequently, cDNA was also regarded as patent eligible. Moore J concurred with the outcome, placing cases like *Parke-Davis* and *In re Marden* along a spectrum where one critical factor was the 'enlargement of the range of utility' of the claimed product compared to its natural counterpart.[73]

The Court rejected both claims to 15 nucleotide sub-sequences, deeming claim 5 'breathtakingly broad.'[74]The cDNA sub-sequence claim also faced scrutiny as it would cover parts of the cDNA of 'more than 4% of human genes,' where Myriad had contributed nothing.[75]The Court suggested that Myriad could have fared better if it had confined its claims to specific sequences with markedly different characteristics from those found in nature, such as tagged sub-sequences or primers.[76]

Following the Federal Circuit decision, the AMP petitioned the Supreme Court for a writ of certiorari, requesting that it takes the case for review. In light of its recent opinion in *Mayo*, the Supreme Court granted the writ, overturned the Federal Circuit's first decision, and ordered the lower court to rehear the case.[77] The Federal Circuit held its ground, ruling again in favour of Myriad.[78]

In 2013, the case was ultimately appealed to the Supreme Court, which referred to Section 101 of the AIA, the implicit exceptions, and the recent *Mayo* decision. The Court's analysis started by distinctly delineating between unpatentable subject matter and patentable subject matter. It reiterated the statutory exceptions, namely laws of nature, natural phenomena, and abstract ideas, affirming that they fall outside the domain of patent protection as they are fundamental tools of scientific and technological work.[79] The Court clearly differentiated the present case with the decision in *Chakrabarty* providing that the bacteria developed by Chakrabarty was non-naturally

[69] *Myriad (Court of Appeals II)*, 689 F 3d 1303, 1327–8 [2012].

[70] Ibid., 1328.

[71] Ibid., 1329.

[72] Ibid., 1330.

[73] Ibid.

[74] Ibid., 1356.

[75] Ibid.

[76] Ibid.

[77] *Association for Molecular Pathology v. Myriad Genetics, Inc.*, 132 S.Ct. 1794 [2012].

[78] *Myriad (Court of Appeals II)*, 689 F 3d 1303, 1327–8 [2012].

[79] *Mayo*, 566 U.S., 132 S.Ct., at 1293.

occurring as it was new 'with markedly different characteristics from any found in nature,' due to the new plasmids that were added to bacteria via recombination.[80] Compared with the present case, the Court concluded that, 'Myriad did not create anything. To be sure, it found an important and useful gene, but separating that gene from its surrounding genetic material is not an act of invention.'[81] The Court further held that it is not sufficient for an invention to be the product of great labour or even deep insight 'groundbreaking, innovative, or even brilliant discovery does not by itself satisfy the Section 101 inquiry.'[82]

The analysis extended to the fundamental issue of conceptualising genes as information or ordinary molecules.[83] Myriad argued that isolated DNA is not naturally occurring and hence patent eligible, as it's 'cut out' from larger naturally occurring DNA. The Court countered that Myriad's claims focused on genetic information encoded, not on chemical composition changes.[84] The Court also ruled that if Myriad had claimed a defined, specific molecule accurately, a hypothetical infringer could have avoided the patent using slightly extended sequences beyond Myriad's claims.[85]

The Supreme Court further analysed the claims directed towards cDNA as 'not naturally occurring,' concluding that for this reason those claims did 'not present the same obstacles to patentability as naturally occurring, isolated DNA segments.'[86] Even if the cDNA sequence was actually 'dictated by nature,' it was still regarded as 'something new,' in the sense of being a 'distinct' and artificial creation.[87] The Court relied on *Chakrabarty* for the proposition that a new product of human ingenuity 'having a distinctive name, character, [and] use' could be patentable subject matter.[88] Adding to the above, the court suggested that where the size of a DNA sequence is so short that it does not contain introns, cDNA 'may be indistinguishable' from natural DNA.[89]

In conclusion, the Court disowned any intention to address or comment on the eligibility of method claims, claims for '*applications* of knowledge about the BRCA1 and BRCA2 genes,' or claims that are directed towards 'alteration of the genetic code.'[90] Thomas J summarised the holding: 'genes and the information they encode are not patent eligible under Section 101 simply because they have been isolated

---

80 *Chakrabarty* 447 U.S., at 310, 100 S.Ct. 2204.

81 *Association for Molecular Pathology v. Myriad Genetics, Inc.*, 569 U.S. 576 [2013].

82 Ibid., 12.

83 *Myriad (District Court)*, 702 F Supp 2d 181, 228 [2010].

84 *Association for Molecular Pathology v. Myriad Genetics, Inc.*, 569 U.S. 576 [2013] at 14.

85 Ibid., 15.

86 Ibid., 16.

87 Ibid., 17.

88 Ibid., 12; *Diamond v Chakrabarty*, [1980] USSC 119; 447 US 303, 309–10 (1980) citing *Hartranft v Wiegmann*, [1898] USSC 78; 121 US 609, 615 (1887).

89 *Association for Molecular Pathology v. Myriad Genetics, Inc.*, 569 U.S. 576 [2013] at 17.

90 Ibid.

from the surrounding genetic material.'[91] However, the Court also held that cDNA is patent eligible because it is not naturally occurring.

Comparing the conclusions of US Supreme Court decision on patenting of genetic material to Australian High Court decision on the same issue shows some similarity and dissimilarity in the approaches in the two jurisdictions. Specifically, as discussed in Sect. 3.1.1 of Chap. 3, the decision delivered by High Court of Australia in the *D'Arcy* case provided restrictions on patentability of genetic material considering that the BRCA1 gene (as claimed) does not demonstrate an artificially created state of affairs. This is similar to the approach that the US Supreme Court took in *AMP* (in light of *Chakrabarty*) that a new product of human ingenuity 'having a distinctive name, character, [and] use' could be patentable subject matter. In contrast to the US Supreme Court's decision in *AMP*, the High Court of Australia in *D'Arcy* did not comment on the patentability of cDNA. However, the Court commented on the consequences of invalidity and held that:

> It was common ground that if claims 1-3 did not contain patentable subject matter, then those claims would not be saved where they extend to forms of nucleic acid that have been synthesised in the laboratory (cDNA).[92]

In light of the principles laid down by High Court of Australia in *D'Arcy*, IP Australia provided specific guidelines on the patentability of nucleic acid and genetic information including cDNA. Specifically, the guidelines provided by IP Australia indicated that,

> …the act of isolation, purification or synthesis is not enough to confer patentability in this situation. This is the case even when the molecule is man-made (for example cDNA) if the genetic information in the man-made molecule is the same as that in the genome of an organism, because the genetic information conveyed by the claimed molecule has not been changed, and therefore cannot be considered 'made' for the assessment of manner of manufacture.[93]

It is evident that the Australian approach on patentability of cDNA is different from the US. However, if a patent application claims genetic information that is different from its natural form it may be considered as 'made' and may satisfy the manner of manufacture requirements in Australia.

The *AMP* decision has implications for bioprinted tissues and bioinks, as they may involve natural products or processes that are derived from or based on DNA sequences. For example, bioinks may contain natural cells or biomolecules that have specific functions or interactions encoded by DNA. Bioprinted tissues may rely on natural processes such as gene expression or regulation that are controlled by DNA. To be patentable, these products or processes must have markedly different characteristics from what occurs in nature and must not claim the natural information or phenomena alone. Additionally, any claims related to bioprinting or bioink preparation must have an inventive concept that adds something significant to the natural product or process.

---

[91] Ibid., 18.

[92] *D'Arcy v Myriad Genetics Inc* [2015] HCA 35 at 283.

[93] See Sect. 3.1.1 of Chap. 3; IP Australia (2020), *Manual of Practice and Procedure*, Part 2.9.2.6.

4.3.5 Cloned Sheep (Dolly)—In re Roslin Institute

In a 2014 decision, *In re Roslin Institute* case (*Roslin*), the Federal Circuit applied the 'markedly different characteristics test' from *Chakrabarty* to reject the subject matter eligibility of cloned sheep 'Dolly.'[94] The subject matter of the patent application ('233 patent application) in issue claims cloned mammals as products which have been rejected by the USPTO and the IPAB as the respective claims are considered unpatentable in light of the provisions of Section 101.[95] Roslin Institute, the assignee of the patent, appealed the decision. Specifically, the '233 application claimed the products of the inventors' unique cloning method: cattle, sheep, pigs, and goats. Claims 155 and 164 are representative:

> 155. A live-born clone of a pre-existing, non-embryonic, donor mammal, wherein the mammal is selected from cattle, sheep, pigs, and goats.
>
> 164. The clone of any of claims 155 — 159, wherein the donor mammal is non-foetal.

The Federal Circuit looked into the claims at issue in light of *Chakrabarty* and *AMP*.[96] Specifically, the Federal Circuit examined the claims with the 'markedly different characteristics test' provided in *Chakrabarty*. The Court in Roslin indicated that the clones claimed in the '233 patent application were identical to their donors. The Court provided that:

> ...Dolly herself is an exact genetic replica of another sheep and does not possess 'markedly different characteristics from any [farm animals] found in nature.'... Dolly's genetic identity to her donor parent renders her unpatentable.[97]

Considering the conclusions of *AMP* (where the Court held that isolated naturally occurring DNA strands are not eligible for patent protection), the Federal Circuit held that:

> ...Roslin 'did not create or alter any of the genetic information' of its claimed clones, '[n]or did [Roslin] create or alter the genetic structure of [the] DNA' used to make its clones.[98] Instead, Roslin's chief innovation was the preservation of the donor DNA such that the clone is an exact copy of the mammal from which the somatic cell was taken. Such a copy is not eligible for patent protection.[99]

The Federal Circuit thus concluded that claims directed towards clones should not be considered patent eligible because they are identical copies of naturally occurring organisms. This case is particularly relevant in the present context as bioprinted

---

[94] *In re Roslin Institute (Edinburgh)*, 750 F.3d 1333, 1334 Fed. Cir. [2014]; Dolly was a female domestic sheep, and the first mammal cloned from an adult somatic cell, using the process of nuclear transfer.

[95] *US Patent No. US09225233*, filed on 04 January 1999 ('Expired' on 03 January 2019).

[96] *Diamond v. Chakrabarty*, 447 U.S. 303, 309 [1980]; *Association for Molecular Pathology v. Myriad Genetics, Inc.*, 569 U.S. 576 [2013].

[97] *In re Roslin Institute (Edinburgh)*, 750 F.3d 1333, 1334 Fed. Cir. [2014], I.

[98] *Association for Molecular Pathology v. Myriad Genetics, Inc.*, 569 U.S. 576 [2013] at 2116.

[99] *In re Roslin Institute (Edinburgh)*, 750 F.3d 1333, 1334 Fed. Cir. [2014].

tissues are developed with the aim of replicating natural tissues. However, they are not identical copies of naturally occurring tissues. An in-depth analysis of this issue will be provided in the upcoming sections.

The *Roslin* case is relevant to the subject matter related to bioprinted tissues and bioinks because it illustrates the application of the 'markedly different characteristics test' from Chakrabarty to determine the patent eligibility of biological products. The Federal Circuit held that clones are not patent eligible because they are exact copies of natural organisms and do not have any distinctive features from them. However, bioprinted tissues and bioinks are not exact copies of natural tissues. They involve modifications of biological materials and structures to create functional and viable products. Therefore, they may have markedly different characteristics from natural tissues and may be eligible for patent protection. However, the extent and nature of these modifications and their impact on the patent eligibility of bioprinted tissues and bioinks are not clear and require further analysis.

4.3.6 Computer-Implemented Scheme—Alice

Shortly after the *Roslin* decision, the US Supreme Court handed down a decision in *Alice Corp. Pty. Ltd. v. CLS Bank Int'l* (*Alice*) where the Court unanimously decided that the claims in issue were not patent eligible under Section 101.[100] A computer-implemented scheme to reduce 'settlement risk' by using a third-party broker was disclosed in the patents in this case.[101] The question raised was whether these claims were patent eligible or were instead nothing more than patent-ineligible abstract ideas. Following *Mayo*, the US Supreme Court reapplied its two-step test for analysing whether or not a claim is patent eligible where the claimed subject matter includes an abstract idea.[102] First, the Court held it was necessary to determine whether or not the claim is 'directed to' an abstract idea. While every claim may be considered to relate in some manner to an abstract concept, this decision appears to indicate that many such claims are not 'directed to' an abstract idea, at least not to an 'abstract idea' as the Court understood that term. Secondly, the Court held it was necessary to determine whether or not the claim contains an 'inventive concept' outside the abstract idea, with the term 'inventive concept' defined to mean 'an element or combination of elements that is "sufficient to ensure that the patent in practice amounts to significantly more than a patent upon the [ineligible concept] itself".'[103]

Specifically, while interpreting Section 101, the Court indicated that: 'We must distinguish between patents that claim the building blocks of human ingenuity and

[100] *Alice Corp. Pty. Ltd. v. CLS Bank Int'l*, 134 S. CT. 2347 [2014].

[101] The patents at issue are United States Patent Nos. 5,970,479 (the '479 patent), 6,912,510, 7,149,720, and 7,725,375.

[102] Ibid.

[103] Ibid.

those that integrate the building blocks into something more, thereby transforming them into a patent-eligible invention.'[104] The Court stated that an abstract idea could not be patented simply because it was executed using a computer, and further, because an escrow arrangement is only an embodiment of an abstract idea, it is not a patentable invention. The Court held that the 'Mayo Framework' should be applied in circumstances where the Court must determine whether a claim is patent eligible or not. The Court reasoned that this 'two-step' methodology must be used to assess whether a patent claimed an abstract idea that is ineligible for patent protection or a possibly patentable practical implementation of an idea.

The implications of this case analysis are particularly relevant to the realm of 3D bioprinting, as it establishes a rigorous standard for patenting computer-implemented inventions that involve abstract ideas. In this context, the term 'bioink,' which refers to the composition of living cells, biomaterials, or active biomolecules employed in 3D bioprinting, could be deemed an abstract idea if it fails to contribute to the technical efficacy of generating tissue-like 3D structures. Furthermore, the bioink must also fulfil the criteria of resolving a technical problem and producing additional technical effects, such as faithfully replicating the properties of natural tissue and facilitating cellular functionalities.

4.3.7 Methods of Preserving Hepatocyte Cells—Rapid

In 2016, the Federal Circuit in *Rapid Litigation Management Ltd v. CellzDirect, Inc.* (*Rapid*) ruled that a method of hepatocyte cryopreservation is patent eligible under Section 101.[105] Hepatocytes are a particular type of liver cells that have several characteristics that are advantageous for testing, diagnosis, and treatment.[106] However, several factors restrict their use such that fresh hepatocytes have a short lifespan and can only be obtained from liver resections or non-transplantable livers of organ donors. Since supply is inconsistent, this means that availability is finite and uncertain.[107] The inventors of the patent US7604929B2 ('929) observed that some hepatocytes could endure many freeze–thaw cycles and remain viable for research and medical applications.[108] Patent '929 included claims directed towards a process of collecting these unique hepatocytes that involved separating viable from non-viable hepatocytes using density gradient fractionation, then refreezing the viable cells from a pool of previously frozen and thawed hepatocytes from various donors.[109]

[104] *Alice Corp. Pty. Ltd. v. CLS Bank Int'l*, 134 S. CT. 2347 [2014], 2354.

[105] *Rapid Litig. Mgmt. v. CellzDirect, Inc.*, 827 F.3d 1042, 1049, 119 USPQ2d 1370, 1374 [2016].

[106] Ibid., I 4.

[107] Ibid.

[108] *US Patent No. US7604929B2*, filed on 21 April 2005 ('Granted' on 20 October 2009).

[109] Ibid.

The question the Federal Circuit addressed in this case was whether the patent claims are directed to a patent-ineligible concept under Section 101.[110]

In addressing this question, the Federal Circuit applied the *Mayo/Alice* test and determined that the fact that a claimant may describe a patent primarily in terms of the 'natural ability of the subject matter,' as in the case of a cancer treatment being described as 'cancer cells' inability to survive chemotherapy,' does not preclude that patent from being eligible under Section 101.[111] The Court noted that the '929 patent did not even try to claim a different variety of resilient hepatocytes. Instead, the '929 patent improved prior art using a natural discovery.[112] The Court then examined whether step 2 of the *Mayo/Alice* test has been satisfied, and, citing *Diehr*, found that the claims 'improve an existing technological process,' [and thus] are sufficient to 'transform the process into an inventive application of the patent-ineligible concept.'[113]

The court added that the ability of hepatocytes to survive numerous freeze–thaw cycles was not all that was claimed by the '929 patent. Instead, a 'method of producing a desired preparation of multi-cryopreserved hepatocytes' was claimed in the '929 patent. For the Court, this new and improved method, which produced a tangible, usable outcome, clearly did not fall under the description of an invention that was 'directed to' patent-ineligible concepts.[114]

The *Rapid* case holds significance for subject matter related to bioprinted tissues and bioinks, as it exemplifies the application of the *Mayo/Alice* test to determine the patent eligibility of biological processes involving natural phenomena. The Federal Circuit ruled that a method of hepatocyte cryopreservation qualifies as patent eligible because it does not merely claim the inherent ability of hepatocytes to survive freeze–thaw cycles, but instead presents a novel and improved method for producing multi-cryopreserved hepatocytes with specific desired properties. This approach enhances an existing technological process and yields tangible, practical results. In a similar vein, bioprinted tissues and bioinks are not mere natural materials but products of innovative processes that alter biological materials and structures to create functional and viable constructs. Consequently, they could potentially meet the requirements for patent eligibility under the *Mayo/Alice* test by demonstrating that they involve inventive applications of natural phenomena rather than being directed towards the phenomena themselves.

[110] *Rapid Litig. Mgmt. v. CellzDirect, Inc.*, 827 F.3d 1042, 1049, 119 USPQ2d 1370, 1374 [2016].

[111] Ibid., II A 10.

[112] Ibid.

[113] Ibid., II B 13.

[114] Ibid.

4.3.8 Methods of Treatment—Vanda

The Federal Circuit ruled in 2018 that the claims at issue in *Vanda Pharmaceuticals Inc. v. West-Ward Pharmaceuticals* (*Vanda*) are patent eligible under Section 101 because they were not 'directed to' a judicial exception.[115] The claims described a technique of treating a patient with schizophrenia using iloperidone, a medication known to cause QTc prolongation in people with a specific genotype linked to poor drug metabolism. The QT interval is the time between the Q and T waves of the heart rhythm. When corrected for the patient's heart rate it is abbreviated QTc.[116] QT prolongation occurs when the heart muscle contracts and relaxes more slowly than usual. Sudden cardiac arrest and the likelihood of developing irregular heart rhythms may both be increased by QT prolongation.[117] In order to treat a particular disease, the primary phases are determining with a genotyping assay and administering a specific dosage of medication depending on that assessment.[118] The Federal Circuit distinguished this case from *Mayo*, noting that:

> The inventors recognized the relationships between iloperidone, CYP2D6 metabolism, and QTc prolongation, but that is not what they claimed. They claimed an application of that relationship. Unlike the claim at issue in Mayo, the claims here require a treating doctor to administer iloperidone.[119]

As the claims in the Vanda's patent were 'directed to a method of using iloperidone to treat schizophrenia' rather than being directed to a judicial exception, the Federal Circuit determined that the claims in the Vanda patent were eligible under the first step of the Alice/Mayo framework.

The *Vanda* ruling by the Federal Circuit serves as a notable example of several key concepts in subject matter eligibility analysis. First, the Federal Circuit determined that the claims were not 'directed to' the alleged natural relationship between the patient's genotype and the risk of QTc prolongation. The Federal Circuit emphasised that the claims must be considered as a whole, including the arguably conventional genotyping and treatment steps. Secondly, in considering the difference between this case and Mayo, the Federal Circuit emphasised 'the distinction between technique of treatment claims and those in *Mayo*.'[120] The Federal Circuit in this case held that the Supreme Court had distinguished method of treatment claims (which apply natural relationships rather than being 'directed to' them) as not being involved by its rulings in Mayo and Myriad because they 'confine their reach to particular applications in light of the previous Mayo and Myriad decisions.'[121] The Federal Circuit emphasised

---

[115] *Vanda Pharmaceuticals Inc. v. West-Ward Pharmaceuticals*, 887 F.3d 1117 [2018].

[116] Ibid., 1121.

[117] Eline Vandael et al., 'Risk factors for QTc-prolongation: systematic review of the evidence' (2017) 39(1) *International journal of clinical pharmacy* 16–25.

[118] *Vanda Pharmaceuticals Inc. v. West-Ward Pharmaceuticals*, 887 F.3d 1117 [2018] at 1134.

[119] Ibid., 1135.

[120] Ibid.

[121] Ibid.

that while the 'claim in *Mayo* recited administering a thiopurine drug to a patient, the claim as a whole was not directed to the application of a drug to treat a particular disease.'[122] In other words, even though the *Mayo* claims mentioned a procedure of administering a patient a medicine, that step was actually carried out to obtain information about the natural relationships and was therefore unrelated to the claims' main diagnostic purpose. The *Mayo* claims did not actually use a natural relationship as a 'method of treatment.'

The *Vanda* case decision holds relevance for subject matter related to bioprinted tissues and bioinks as it offers valuable guidance on how to structure claims for treatment methods involving natural relationships to avoid Section 101 rejections. The Federal Circuit highlights the crucial factor of directing claims towards specific applications of natural relationships, rather than focusing on the relationships themselves. For instance, a claim describing a method of bioprinting a tissue construct using a patient-specific bioink containing tailored cells and biomolecules based on the patient's genotype and phenotype may be deemed patent eligible. This is because it applies the natural relationship between the patient's genetic and physical characteristics and the optimal bioink composition to treat a particular disease or condition. On the other hand, a claim that merely recites the natural relationship between the patient's genotype and phenotype and the bioink composition without detailing the bioprinting process or treatment approach may face rejection as a judicial exception.

4.3.9 Diagnostic Methods—Illumina

In *Illumina, Inc. v. Ariosa Diagnostics, Inc.* (Illumina), the Federal Circuit ruled that patents claiming methods for creating an extracellular fraction of cell-free DNA that is enriched in foetal DNA were patent eligible and not invalid under Section 101.[123] Specifically, the methods of creating a portion of cell-free DNA that was enriched in foetal DNA were the subject of the claims in patents in suit. Earlier, the prior art found that maternal plasma contained circulatory extracellular foetal DNA as well as circulatory extracellular maternal DNA. The prior art further disclosed that foetal circulatory extracellular DNA was relatively small, with a size of 500 base pairs or less, compared to maternal circulatory extracellular DNA, which has a size greater than 500 base pairs. The present invention was based on this size disparity and could be employed to separate and enrich for foetal DNA sequences from the circulatory extracellular maternal DNA. The enriched foetal DNA sequences could then be evaluated for abnormal genetic loci in the foetal DNA.[124]

Considering the current state of the law on Section 101 patentability, the majority appeared to take a rather restrictive and formalistic approach. The Court considered a 'bucket approach' in analysing the claims in issue and indicated that if the claims

[122] Ibid., 1134.

[123] *Illumina, Inc. v. Ariosa Diagnostics, Inc.* 952 F.3d 1367 [2020].

[124] Ibid., 1369, 1370.

fit into a specific 'bucket,' that will determine whether they are patent eligible. The claims would not be patent eligible if they fall under the category of 'diagnostic claims.' However, if the claims fit under the category of 'method of treatment,' then the claims would satisfy Section 101.[125] The claims in *Illumina* did not fit into either of the 'buckets' established by *Mayo* and *Alice*, thus the Court proposed a new category of patent eligibility for claims related to methods of preparing biological samples.[126]

The majority concluded that 'cell-free foetal DNA tends to be shorter than cell-free maternal DNA in a mother's bloodstream' and this was the natural phenomenon that the inventors had discovered which allowed the essential cell-free foetal DNA ('cffDNA') to be analysed for specific birth defects instead of pregnant women needing to undergo riskier procedures.[127] This conclusion was made with regard to the first step of determining patent eligibility under *Mayo/Alice*. The Court determined that because the claims in this case were directed to 'methods for preparing a fraction of cell-free DNA that is enriched in foetal DNA,' they were not 'directed' to the natural phenomenon, such as the cffDNA or its size difference.[128]

By establishing strict limits between patent-eligible claims (method of treatment claims and method of preparation claims) and ineligible claims, the majority's 'buckets' approach to addressing Section 101 issues in subject matter related to life sciences appeared to offer some clarity to the present subject matter eligibility scenario. According to the majority, in order to avoid being classified ineligible for claiming a natural phenomenon under Section 101, patent claims with subject matter related to life sciences such as methods to diagnose or methods to detect disease could be changed to methods of preparing samples or even methods of treatment.

A comparative analysis of the conclusions of *Illumina* case reveals interesting insights as the same patent applications were the subject of the litigation analysed in Sect. 3.2.2 of Chap. 3.[129] As discussed above, the Federal Circuit in *Illumina* held that the claims are not directed to the natural phenomenon that cell-free foetal DNA is generally shorter than cell-free maternal DNA but instead are directed to 'a patent-eligible method that utilises it.' Similarly, the Full Court of Australia considered that the claims in issue are patent eligible by reasoning that the claims in the *Sequenom* patent were directed towards an artificially created state of affairs that resulted in an economically useful outcome.[130] Despite differences in reasoning, it is important to note that the US and Australian courts produced similar outcomes. It is evident that patent applications with claims directed towards diagnostic methods utilising natural phenomena or products of nature would be considered patentable in the US and Australia.

---

[125] Ibid., 1371, 1372.

[126] Ibid., 1377.

[127] Ibid., 1375.

[128] Ibid.

[129] See Sect. 3.2.2 of Chap. 3.

[130] *Ariosa Diagnostics Inc v Sequenom Inc* [2021] FCAFC 101 at 159.

4.4 USPTO Guidelines on Patentable Subject Matter Eligibility

The USPTO provides precise guidelines on the requirements for satisfying the patentable subject matter requirement. In particular, the USPTO has provided two criteria for subject matter eligibility that are based on the analysis of the Supreme Court in *Alice* and *Mayo.*[131] According to the first criterion, the subject matter claimed must be in one of the four statutory categories as described in Section 101. The four categories of invention considered by Congress to be acceptable subject matter include processes, machines, manufactures, and composition of matter. Specifically, 'things' or 'products' are defined in the last three categories, while 'actions' are defined in the first category (i.e., inventions that consist of a series of steps or acts to be performed).[132]

Secondly, the claimed subject matter must qualify as subject matter which is patent eligible, that is, the patent must not claim a judicial exception unless the claim as a whole involves additional elements which are considerably greater than the exception. Judicial exceptions (also referred to as 'judicially recognised exceptions' or merely 'exceptions') are the subject matter which the courts have considered to be beyond, or exceptions to, the four formal classes of invention that are limited to abstract ideas, laws of nature, and natural phenomena (including products of nature).[133]

The examination of subject matter by the USPTO relies on the foundations laid by the US Courts in *Alice* and *Mayo.* Specifically, the examination guidelines detail that the only test that ought to be applied to determine whether claims under consideration are eligible is the *Alice/Mayo* two-part test. The examiners are not required to employ the machine-or-transformation test as a stand-alone eligibility test, despite this test being a crucial indicator of eligibility. Instead, it ought to be taken into account as a component of the 'integration' or 'substantially more' determination described in the *Alice/Mayo* test.[134]

Examiners at the USPTO would apply the same two-step procedure to assess the subject matter eligibility of bioprinted tissues and bioinks, just as they do with other types of patent applications. In the following sections of this chapter, I will examine product and process claims related to bioinks and bioprinted tissues, considering

---

[131] United States Patent and Trademarks Office (2019), *Manual of Patent Examining Procedure*, Part 2106.

[132] See 35 U.S.C. 100(b); ('The term "process" means process, art, or method, and includes a new use of a known process, machine, manufacture, composition of matter, or material.'); United States Patent and Trademarks Office (2019), *Manual of Patent Examining Procedure*, Part 2106.03.

[133] United States Patent and Trademarks Office (2019), *Manual of Patent Examining Procedure*, Part 2106.04; *Alice Corp. Pty. Ltd. v. CLS Bank Int'l*, 573 U.S. 208, 216, 110 USPQ2d 1976, 1980 (2014) (citing *Ass'n for Molecular Pathology v. Myriad Genetics, Inc.*, 569 U.S. 576, 589, 106 USPQ2d 1972, 1979 (2013).

[134] United States Patent and Trademarks Office (2019), *Manual of Patent Examining Procedure*, Part 2106.04(d), 2106.05(b), (c).

the requirements of patentable subject matter under US law. This analysis will also involve a comparative study of the conclusions drawn in Chap. 3.

4.5 What Is the Impact on Patentability of Bioprinted Tissues and Bioinks?

4.5.1 Bioprinted Tissues and Bioinks: Product Claims

Patent claims directed towards bioinks can be considered as 'compositions of matter' as the bioinks are composed of living cells, growth factors, biopolymer gel, and other elements that are useful for the stabilisation and further proliferation of living cells.[135] Similarly, patent claims directed towards bioprinted tissues as products can be considered 'products of manufacture' and/or 'compositions of matter' as bioprinted tissues are end products made using bioinks and other elements that act as structural building blocks of the bioprinted tissues.[136] Thus, patent claims directed towards bioinks and bioprinted tissues may overcome the initial screening of 'patentable subject matter' on the basis that they fall within the scope of 'products of manufacture' and/or 'compositions of matter.' However, in accordance with the *Mayo/Alice* test, the subject matter may only be considered patentable (according to Section 101) if the claimed subject matter is not within the scope of judicial exceptions—abstract ideas, laws of nature, and natural phenomena (including products of nature). By way of comparison, under Australian law, regulators are more focused on identifying whether the subject matter related to bioprinted tissues or bioinks as products satisfies the manner of manufacture requirements and may be considered as artificially created state of affairs.

As living cells and other active molecules make up bioinks and bioprinted tissues, it is crucial to determine whether a specific subject matter falls under the purview of the available judicial exceptions, particularly the 'products of nature' exception. The outcome of this analysis may differ from case to case, depending on how the subject matter related to bioinks and bioprinted tissues is claimed. For example, if a patent claim is directed towards a bioink that is composed of living cells, growth factors, biopolymer gel, and other elements in a way which is not similar to how the living cells exist in a natural state, specific claim may not fall within the scope of 'products of nature' and may be considered patentable.

The current provisions are helpful in accessing the subject matter in light of the analysis of patentable subject matter provisions with respect to bioinks and bioprinted tissues as products in this section. However, there is some uncertainty with regards to

---

[135] Timo Minssen and Marc Mimler, 'Patenting Bioprinting-Technologies in the US and Europe—The 5th Element in the 3rd Dimension' (2016).

[136] Sean V. Murphy and Anthony Atala, '3D bioprinting of tissues and organs' (2014) 32(8) *Nature biotechnology* 773–785.

specific provisions such as 'products of nature' and 'human organism.'[137] US courts have attempted to address uncertainty in the provisions relating to patentable subject matter requirements and judicial exceptions, and it is always important to analyse the subject matter in light of recent decisions as individual decisions establish precedents that may give an appropriate direction to understand and apply the relevant laws. Hence, further sections of the chapter will focus on analysis of the patentable subject matter issues surrounding bioinks and bioprinted tissues in light of the relevant US court decisions.

Considering the analysis of Section 101 in the *Chakrabarty* case, it is clear that a non-natural creation or composition of matter and a product of human ingenuity is covered within the language of Section 101. This is relevant to the present analysis of product claims directed towards bioinks and bioprinted tissues. Although the composition of bioinks varies on a case-by-case basis, bioinks in general are composed of living cells and other natural and synthetic elements. A bioink composition consisting of naturally occurring or naturally available elements and synthetic elements may be considered patent eligible considering the conclusions of *Chakrabarty* case and the provisions provided in Section 101.

Similarly, bioprinted tissues are composed of bioink, bioactive molecules, and growth factors that support the proliferation and development of the bioprinted tissues.[138] While the composition of bioprinted tissues may also vary on a case-to-case basis, *Chakrabarty* emphasises that, to be considered patent eligible, a claim directed towards a bioprinted tissue as a whole should not be naturally occurring and should be a product of human ingenuity. For example, a patent application with a claim directed towards 'an engineered, living, three-dimensional liver tissue construct' has recently been granted by USPTO.[139] The claim reads:

> An engineered, living, three-dimensional liver tissue construct comprising:
>
> (a) at least one layer comprising a plurality of cell types spatially arranged relative to each other in a planar geometry; or
>
> (b) a plurality of layers, at least one layer compositionally or architecturally distinct from at least one other layer in a laminar geometry, wherein the at least one other layer comprises a plurality of cell types spatially arranged relative to each other in a planar geometry;
>
> wherein the plurality of cell types comprises parenchymal cells in contact with non-parenchymal cells in at least one compartment that comprises an interior comprising the parenchymal cells and a border comprising the non-parenchymal cells,
>
> wherein the parenchymal cells comprise hepatocytes or hepatocyte-like cells and the non-parenchymal cells comprise endothelial cells, and
>
> wherein the construct is substantially free of pre-formed scaffold and adherent confinement material at the time of use in an assay, implantation into a subject, or use in extracorporeal support of a subject.

[137] Jun-Long Chen, 'Patent Eligibility Analysis of Bioprint Technology' (2017) 6(1) *NTUT Journal of Intellectual Property Law and Management* 46–66.

[138] Clarissa Tomasina et al., 'Bioprinting vasculature: materials, cells and emergent techniques' (2019) 12(17) *Materials* 2701.

[139] *US Patent No. US10400219B2*, filed on 26 May 2016 ('Granted' on 03 September 2019).

The preamble of the above provided claim describes the claimed invention as 'an engineered, living, three-dimensional liver tissue construct....' Regarding the interpretation of claim language, the MPEP provides that 'the words of a claim must be given their "plain meaning" unless such meaning is inconsistent with the specification.'[140] The plain English meaning of the word 'engineered' is 'skilfully and deliberately arranged rather than arising naturally or spontaneously.'[141] According to this plain English meaning, the living, three-dimensional liver tissue construct as claimed in the above provided claim would be non-naturally occurring. Additionally, the geometrical arrangement of cells in multiple layers is distinct from each other, each layer comprising different cell types. The basic composition including the type of cells selected for bioprinting bear similarities to the cells of natural liver tissue to ensure the bioprinted liver tissue possesses the properties of natural liver tissue. However, the structural composition of bioprinted liver tissue as a whole is different from naturally available liver tissue and on the face of it should be considered patent eligible.[142]

The Supreme Court's findings in *AMP* highlight that patent eligibility for bioprinted tissues and bioinks necessitate the inclusion of subject matter that is artificially created. While these materials mimic natural tissue architecture and incorporate natural elements like stem cells and molecules, their entirety is non-naturally developed through human efforts and labour.[143] The distinction made by the Court between naturally occurring gene sequences and patent-eligible non-naturally occurring cDNA underscores the importance of demonstrating an artificial aspect in the subject matter to qualify for patent protection. Thus, bioinks and bioprinted tissues, as a whole, can be seen as products resulting from human intervention and may be eligible for patent protection.

The decision of the Federal Circuit in *Roslin* raises questions as to the patent eligibility of product claims directed towards bioinks and bioprinted tissues. In *Roslin*, the Federal Circuit clearly indicated that Dolly is an exact copy (genetically and physically) of a naturally occurring donor parent and does not possess any markedly different characteristics in comparison with the donor parent. Hence the claims directed towards a cloned animal are unpatentable in light of Section 101.[144] The product claims directed towards bioinks and bioprinted tissues may not fall within the scope of this decision as bioinks and bioprinted tissues (as a whole) are not naturally occurring but are composed of products of nature such as living stem cells and bioactive molecules *as well as* synthetic molecules such as growth factors that support the growth of living stem cells. These additional elements are key. It is

---

[140] United States Patent and Trademarks Office (2019), *Manual of Patent Examining Procedure*, Part 2111.01.

[141] 'Engineered', *Lexico powered by Oxford*, https://www.lexico.com/definition/engineered.

[142] Clark Amanda et al., 'A microphysiological system model of therapy for liver micrometastases' (2014) *Experimental biology and medicine* 239.9: 1170–1179.

[143] Anni Sorkio et al., 'Human Stem Cell Based Corneal Tissue Mimicking Structures using Laser-Assisted 3D Bioprinting and Functional Bioinks' (2018) 171 *Biomaterials* 57–71.

[144] *In re Roslin Institute (Edinburgh)*, 750 F.3d 1333, 1334 Fed. Cir. [2014].

submitted that bioinks and bioprinted tissues may be considered to have markedly different characteristics in comparison with the composition of tissue available in nature. Similarly, the structural configuration and composition of natural tissues is complex and not identical to the structural configuration and composition of man-made bioinks and bioprinted tissues. Hence, bioinks and bioprinted tissues should not be considered to be exact copies of tissues available in nature and must be considered patent eligible in light of Section 101.[145]

Section 33(a) of the AIA provides that patent claims directed towards a 'human organism' are not patentable. Specifically, the legislation states that 'Notwithstanding any other provision of law, no patent may issue on a claim directed to or encompassing a human organism.'[146] The plain English meaning of the word human is 'of, relating to, or characteristic of humans' and the plain English meaning of the word organism is 'an individual constituted to carry on the activities of life by means of parts or organs more or less separate in function but mutually dependent.'[147] Considering the plain English meanings of the term 'human organism,' a broad interpretation might include 'an individual of, relating to, or having characteristic of humans who is constituted to carry on the activities of life.' This suggests that the exception provided by Section 33(a) of the AIA is directed towards a 'human organism' as a whole. In addition to the ethical considerations, the reasoning behind this restriction on patentability is most probably not to provide any monopoly on anything in a patent claim which is directed towards or encompassing a human organism 'as a whole.'[148] Bioprinting technology is currently capable of developing different parts of humans at tissue and organ level in a living form.[149] The technology is not capable of developing a human organism 'as a whole.' Hence, product claims for 3D bioprinting subject matter in its current technological state would be extremely unlikely to fall within the scope of the 'human organism' exception.

In the US, subject matter directed at bioprinted tissues and bioinks may only be regarded patentable if a specific claim falls within one of the four statutory categories as defined in Section 101, and has 'markedly different characteristics' in light of any judicial exception. The judicial exception that is most closely relevant to the topic of bioprinted tissues and bioinks is natural phenomenon or products or nature, since bioprinted tissues or bioinks contain natural elements, such as living cells.

[145] Robert M. Skirvin, 'Natural and induced variation in tissue culture' (1978) 27(1) *Euphytica* 241–266; Ahu Arslan-Yildiz et al., 'Towards artificial tissue models: past, present, and future of 3D bioprinting' (2016) 8(1) *Biofabrication* 014103.

[146] *Leahy-Smith America Invents Act 2011* (US) s 33(a).

[147] 'Human', *Merriam-webster*, https://www.merriam-webster.com/dictionary/human; 'Organism', *Merriam-webster*, https://www.merriam-webster.com/dictionary/organism.

[148] Morten Walløe Tvedt and Ellen-Marie Forsberg, 'The room for ethical considerations in patent law applied to biotechnology' (2017) 20(5–6) *The Journal of World Intellectual Property* 160–177.

[149] Madhuri Dey and Ibrahim T. Ozbolat, '3D bioprinting of cells, tissues and organs' (2020); Nilesh Bhamare et al., '3D printing of human ear pinna using cartilage specific ink' (2021) 16(5) *Biomedical Materials* 055008.

4.5.2 *Bioprinted Tissues and Bioinks: Process or Method Claims*

Considering the Supreme Court's analysis and decision in *Alice/Mayo* with reference to bioinks and bioprinted tissues, it is important to understand that process or method claims directed towards bioinks and bioprinted tissues may not be considered patentable if the process claim encompasses laws of nature alone or would involve a well-understood, routine, conventional process or method of developing or delivering bioinks or bioprinted tissues previously engaged in by researchers in the field. Process or method claims in the majority of patent applications relating to bioink or bioprinted tissues may not directly or indirectly encompass laws of nature in a way that claims a specific law of nature as a whole. This is because the direct involvement of laws of nature is comparatively less important in the process or method of developing or delivering bioinks or bioprinted tissues. Such claims are, on the face of it, seemingly more likely to be upheld. For example, a patent claiming 'a method of printing a tubular tissue construct' has been granted by the USPTO.[150] The claims read as:

> A method of printing a tubular tissue construct comprising:
>
> depositing one or more sacrificial filaments on or in a substrate to form a functional channel pattern, each sacrificial filament comprising a fugitive ink and a plurality of predetermined types of viable cells, wherein each predetermined type of viable cells is deposited at a different predetermined location along a length of the sacrificial filament;
>
> at least partially surrounding the functional channel pattern with an extracellular matrix composition,
>
> removing the fugitive ink to create one or more functional channels in the extracellular matrix composition, at least a portion of each different predetermined type of viable cells remaining at the different predetermined location after removal of the fugitive ink, thereby forming a tubular tissue construct.

The above claim is directed towards a method of bioprinting a tissue. This method claim did not specifically claim any laws of nature and hence passes the *Alice/Mayo* test. Similarly, when considering the markedly different characteristics examination for a process claim, the focus of examination is on the claimed process rather than the elements (may include products of nature) claimed in the process. The evaluation of a process claim should concentrate on the active steps in the process rather than the products of nature used in those steps.[151] This is in line with the Court's analysis in *Rapid*, where a claimed hepatocyte cryopreservation mechanism, consisting of performing density gradient fractionation to distinguish viable and non-viable hepatocytes, recovering viable hepatocytes, and cryopreserving the recovered viable hepatocytes, was evaluated by the Court. The Court did not apply the natural products

---

[150] *US Patent No. US10702630B2*, filed on 04 May 2016 ('Granted' on 07 July 2020).

[151] United States Patent and Trademarks Office (2019), *Manual of Patent Examining Procedure*, Part 2106.04(c).

used in the process to a markedly different characteristics evaluation.[152] Similarly, the example provided above consists of different products of nature, such as viable cells and extracellular matrix composition. These products exist in the same form as they do in nature; however, a markedly different characteristics evaluation was not applied even though the process utilises products of nature. The evaluation of a process claim should concentrate on the active steps in the process rather than the products of nature used in those steps.

However, in a limited case where a process claim relating to a product based on nature is drafted in such a manner that there is little difference between the natural product and the claimed product, the claim will be subject to a markedly different examination of the product.[153] Comparatively, based on the findings of Illumina, where it was determined that the claims to detect naturally occurring cffDNA in maternal blood were addressed to the cffDNA and that 'identifying the presence of cffDNA was merely asserting the natural occurrence itself.'[154] These types of claims are drafted in a manner that focuses on the product rather than the steps involved in the process. For example, consider a claim that recites, in its entirety, 'A method of providing an organ to a human or animal in need thereof.'[155] This claim focuses on the organ itself, which is a product of nature, and the application of the markedly different characteristics examination would identify whether the organ as claimed is markedly different from a naturally existing organ. If the organ is identified to have markedly different characteristics in comparison to its natural counterpart, the claim may satisfy the second step of *Alice/Mayo* and may be considered to be patentable subject matter. But if the organ is identified as having no markedly different characteristics in comparison to an identical natural organ, the claim may not satisfy the second step of *Alice/Mayo*. In essence, if a claim refers to a bioprinted tissue produced using a bioprinting method, the key question is whether the bioprinted tissue produced using a bioprinting method has 'markedly different characteristics' from its counterpart, a natural tissue. In this case, having 'markedly different characteristics' would be understood to mean having additional elements that significantly alter the tissue from its natural state.

---

152 *Rapid Litig. Mgmt. v. CellzDirect, Inc.*, 827 F.3d 1042, 1049, 119 USPQ2d 1370, 1374 (Fed. Cir. 2016).

153 United States Patent and Trademarks Office (2019), *Manual of Patent Examining Procedure*, Part 2106.04(c).

154 *Illumina, Inc. v. Ariosa Diagnostics, Inc.* 952 F.3d 1367 [2020].

155 *US Patent No. US20170368225A1*, filed on 18 December 2015 ('Granted' on 09 June 2020).

4.6 Recommendations: Patentability Framework Model—US

Patent applications with claims directed towards bioprinted tissues and bioinks as products may only be considered patentable if the specific claim falls within one of the four statutory categories as described in Section 101 and has 'markedly different characteristics' pursuant to any judicial exception.[156] As bioprinted tissues or bioinks include natural elements such as living cells, the judicial exception that is closely relevant is natural phenomenon or products or nature.[157] For example, if a claim directed towards a bioink only includes elements that are considered to be products of nature, such a claim may fail the first part (Step 2A) of the *Alice*/*Mayo* two-part test and may not be considered patentable. However, if a similar claim includes additional elements that amount to significantly more than a judicial exception, such a claim may pass the *Alice*/*Mayo* test and would be considered patentable.[158] Similarly, for a claim that is directed to a bioprinted tissue produced by a bioprinting method, the analysis turns on whether the product of nature provided in the claim has 'markedly different characteristics' from its naturally occurring counterpart which is a natural tissue.[159] Here, having 'markedly different characteristics' would be considered as having additional elements that amount to significantly more than a natural tissue which is a product of nature.[160] Similarly, the having 'markedly different characteristics' evaluation of a method claim should concentrate on the active steps in the process rather than the products of nature used in those steps.[161]

A patent application with claims directed towards medical treatment methods that are used for treating a patient (including surgical treatment) would be considered patentable in the US. However, 35 USC 287(c)(1) of AIA refuses a remedy for the infringement of such patented processes or medical activity practised by medical practitioners, effectively nullifying the right in the absence of enforcement.[162] In view of this, a patent application with claims directed towards a method of handheld bioprinting used to bioprint a tissue or bioink as an aspect of treating a patient (including surgically) would be patentable. However, even if such a patent application has satisfied the subject matter and other patentability criteria such as novelty and obviousness, the granted patent may not be enforceable due to the provisions of 35 USC 287(c)(1) of AIA.[163] Utilising this analysis, a patentability framework model has been developed and visualised in Fig. 4.1 to assist in identifying the patentability

[156] See Sect. 4.5.1 of this chapter.

[157] Ibid.

[158] Ibid.

[159] See Sect. 4.5.2 of this chapter.

[160] Ibid.

[161] Ibid.

[162] *Leahy-Smith America Invents Act 2011* (US) s 287(c)(1).

[163] Ibid.

of different types of patent claims directed towards bioprinted tissues and bioinks in the US.

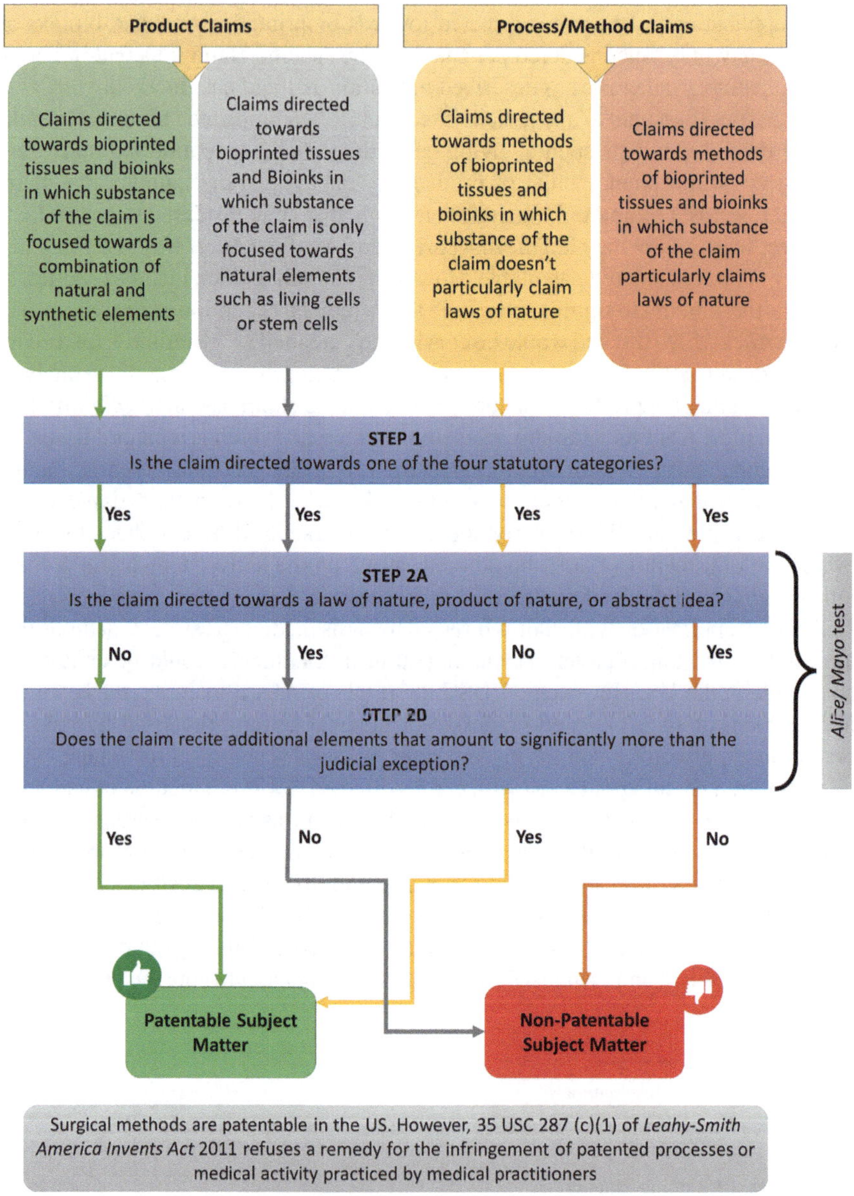

Fig. 4.1 Visualisation of proposed patentability framework model for patenting inventions related to bioprinted tissues and bioinks in the US

To reiterate the central conclusions highlighted in Fig. 4.1, patent applications with product or process claims directed towards bioprinted tissues and bioinks may be considered patentable if they fall within one of the four statutory categories (processes, machines, manufactures, and composition of matter) of invention and satisfy the requirements of *Alice/Mayo* test. Subject matter eligibility may also be contingent on how specific subject matter related to bioprinted tissues and bioinks is claimed by patent applicants. This is supported by the empirical analysis in Chap. 6. A number of recommendations are therefore presented to assist patent applicants in streamlining the application process.

Recommendation 2 (US)

The following strategies are recommended to patent applicants to satisfy the patentable subject matter requirements for inventions with claims directed towards bioprinted tissues or bioinks.

1. A patent application with claims directed towards a bioprinted tissue may be considered to have 'markedly different characteristics' if such a claim includes a combination of natural and synthetic elements.
2. Integration of a combination of natural and synthetic elements would lead to a bioprinting product having 'markedly different characteristics' in comparison to its natural counterpart.
3. Recommendation 2(i) and 2(ii) is also applicable for patent applications with claims directed towards methods of bioprinting. However, for a method claim, the focus should be on the claimed bioprinting process rather than the elements (which may include products of nature) claimed to result from the process.
4. A patent claim directed towards a method of bioprinting a tissue or a bioink directly on the injured site of a human or animal body may be regarded as a method of medical treatment but may not receive any patentable subject matter objection in the US.

4.7 Conclusion

The US patent system's subject matter requirements, outlined in Section 101, have undergone extensive analysis by various courts to offer clarity on their application to different types of subject matter. The USPTO's patent examination guidelines provide valuable guidance, with the *Chakrabarty* decision serving as a landmark example, establishing the patentability of natural elements with markedly distinct characteristics. Similarly, the *Mayo* and *Alice* decisions have introduced a two-step framework for evaluating patent claims involving or related to natural phenomena or products of nature. These rulings significantly influence the approach of US patent examiners when dealing with subject matter associated with natural phenomena or products of nature.

The *Alice/Mayo* two-part test can be used to determine if a subject matter relating to bioprinted tissues or bioinks is patentable. Patent applications that include claims for products such as bioprinted tissues and bioinks may only be considered patentable if a particular claim relates to one of the four statutory categories listed in Section 101 and differs significantly from the others in light of any judicial exception. The judicial exception that is most closely related to bioprinted tissues and bioinks is natural phenomena or products of nature as bioprinted tissues or bioinks comprise natural elements such as living cells. A patent application with claims directed towards bioprinted tissues or bioinks must be drafted in a way that doesn't directly claim a natural phenomenon or product of nature. For example, if a claim directed towards a bioink only includes elements that are considered to be products of nature, such a claim may satisfy the second part of *Alice/Mayo* two-part test. Considering the Courts conclusions in *Rapid, Vanda, and Illumina,* for a claim that is directed to a bioprinted tissue produced by a bioprinting method, the analysis turns on whether the product of nature provided in the claim has markedly different characteristics from its naturally occurring counterpart which is a natural tissue. Despite different challenges, if a patent application with products or processes claims directed towards bioprinted tissues or bioinks can satisfy the *Alice/Mayo* two-part test, such claims would be considered satisfying the requirement of Section 101 and would be considered patent eligible in the US. However, even if a patent application with claims directed towards bioprinted tissues or bioinks satisfies the requirements of Section 101, such an application may need to overcome other patentability criteria such as novelty, inventive step, utility, and disclosure requirements to be considered patentable.

As most of the bioprinted tissues are aimed to be developed in way that mimic natural tissues, it is challenging for scientists and researchers to develop a bioprinted tissue that mimics its natural counterpart as well as exhibiting markedly different characteristics to be considered patent eligible. The practical application of bioprinted tissues with such distinct features adds further complexity. For instance, a bioprinted skin tissue designed for transplantation with markedly different characteristics from natural skin might be considered patent eligible. However, if it proves unsuitable for transplantation due to potential graft rejection, its eligibility could be questionable. The current patentable subject matter requirements, as outlined by US courts and USPTO's examination guidelines, require further clarification in the context of bioprinted tissues and bioinks. To address these uncertainties, Chap. 6 of this book offers an empirical analysis of patent prosecution data pertaining to bioprinted tissues and bioinks, aiming to comprehensively answer the research question from both doctrinal and empirical perspectives.

Chapter 5
Patentable Subject Matter Assessment of 3D Bioprinted Tissues and Bioinks: European Perspective

5.1 Introduction

The subject matter requirements for patent applications with claims related to bioprinted tissues and bioinks show both similarities and differences across three major patent jurisdictions: Australia, the US, and Europe. Chapters 3 and 4 of this book thoroughly examined the patentable subject matter requirements in Australia and the US, providing insights into how their existing patent systems and doctrinal limitations apply to patent applications focused on bioprinted tissues and bioinks. These analyses collectively address the central research question: 'Is the patentable subject matter requirement a hurdle to patenting bioprinting inventions, specifically bioinks and bioprinted tissues?' The findings in Chap. 3 suggest that bioprinting inventions may meet the patentable subject matter requirements under the Australian Patents Act Section 18(1). However, claims solely directed towards natural elements, like living cells or stem cells, might not be considered patentable due to the absence of 'human action' and failure to create an 'artificially created state of affairs.' In contrast, the conclusions in Chap. 4 indicate that patent applications for bioprinted tissues and bioinks as products may be eligible if they fall under one of the four statutory categories described in Section 101 and exhibit 'markedly different characteristics' in light of any judicial exception in the US.

In line with the patentable subject matter analysis of Chaps. 3 and 4 of this book, this chapter will analyse the patentable subject matter provisions of the European Patent Convention (EPC) and relevant legislative bodies of the European Union (EU) and contribute towards answering the research questions of this book.[1] Specifically, the initial sections of this chapter provide an overview of the impact of 3D bioprinting in Europe and subsequent sections discuss the evolution of European patent system

[1] *European Patent Convention*, opened for signature 5 October 1973, (entered in force 7 October 1977); The European Union has an independent executive branch, legislature, and judiciary. These are supported and supplemented by a collection of institutions and bodies whose powers are granted by the foundational Treaties.

P. Devarapalli, *Bioinked Boundaries*, https://doi.org/10.1007/978-981-96-2591-8_5

with a specific focus on how the EPC and EU directives operate. The later sections focus on analysing the relevant legislation and case law related to the patentable subject matter requirements of the EPC and provide analysis on their applicability to patentability of subject matter related to bioprinted tissues and bioinks as 'products' and 'process or methods.' The later section would delve into the Biotechnology Directive's provisions concerning the patentability of bioprinted tissues and bioinks. While focusing on the EPC and EU regulations, this chapter also reflects on the patentable subject matter analyses from Chaps. 3 and 4, which pertain to Australia and the US, respectively. By doing so, this chapter constructs a comparative analysis, highlighting the similarities and differences among these three jurisdictions in regards to the patentability of bioprinted tissues and bioinks. The aim of this chapter is thus to address two of the key research questions of this book, 'Is the patentable subject matter requirement a hurdle to patenting bioprinting inventions, specifically bioinks and bioprinted tissues?' and 'How are the subject matter requirements, and specifically the doctrinal limits as provided by Australian, US, and the EU patent laws and jurisprudence, likely to be applicable to subject matter related to 3D bioprinted tissues and bioinks?'.

As discussed in the previous chapters, 3D bioprinting is a technology that raises a number of patent-related issues.[2] This is because inventions related to 3D bioprinting technologies (specifically bioprinted tissues and bioinks) are complex in nature and encompass a combination of natural and artificial elements.[3] The complexity of bioprinting-related subject matter has triggered a debate on the patentability of such inventions under European law.[4] In light of the expanding commercial applications of bioprinting technologies in Europe and the increase in patenting interests, it is crucial to consider whether the patentable subject matter requirement presents a hurdle to the grant of patents for bioprinting inventions, specifically bioinks and bioprinted tissues, in the context of the EU patent system.

5.2 How Did the European Patent System Evolve?

The European patent system consists of various legislative instruments, including national patent laws, the Strasbourg Convention of 1963, the European Patent Convention (EPC), and EU directives and regulations.[5] The idea of establishing a European Patent Office (EPO) was proposed in 1949 by French Senator Henri Longchambon to the Council of Europe.[6] Although his concept was considered

[2] See Sect. 1.2 of Chap. 1.

[3] See Sect. 2.3 of Chap. 2.

[4] Nabeel M. Althabhawi and Zinatul Ashiqin Zainol, 'The Patent Eligibility of 3D Bioprinting: Towards a New Version of Living Inventions' Patentability' (2022) 12(1) *Biomolecules* 124.

[5] Catherine Seville, *EU intellectual property law and policy* (Edward Elgar Publishing, 2009).

[6] Recommendation 23 of the First Session of the Council of Europe (September 8, 1949, Doc 110). The document Report presented by M.H. Longchambon on behalf of the Committee for Economic Questions offered arguments in favour of the proposal (COE Document No 75, September 3, 1949).

impractical by the Council's Patent Experts Committee, it led to the development of a European patent system.[7] The Council's Patent Experts Committee held meetings that resulted in two Conventions: one on patent application procedures in 1953 and another on the international classification of patents in 1954.[8] The Committee continued its work on substantive patent law, leading to the signing of the Strasbourg Patent Convention in 1963.[9] This multilateral convention aimed to harmonise patent laws among Council of Europe member states and came into force in 1980.[10]

The Munich Diplomatic Conference took place in 1973, leading to the signing of the Munich Convention the following year.[11] Sixteen countries signed the Convention, which came into effect between 1977 and 1978 for various countries. The Munich Convention, also known as the European Patent Convention, established a system for the grant of patents in Europe.[12] European patent law is primarily uniform among parties that have signed the Strasbourg Convention, the EPC, and the TRIPS Agreement.[13] It's worth noting that the EPO is not solely an EU institution, as it includes signatories from non-EU countries like Morocco and Cambodia.[14] The EPC is signed by all EU member states and several other European countries, such as Switzerland, Norway, Hungary, and Bulgaria.[15]

The EPC provides a process for converting national patent applications in signatory nations, granting equal intellectual property rights to applicants across all EPC members.[16] Additionally, applicants seeking patent protection in a specific

---

Only a small number of the COE documents with the 'CM' designation had been digitised at the time of writing; however, catalogue entries for the documents are available on the COE archives' website.

[7] Observations of Governments or Experts on the Draft Proposal for a Convention for the Creation of a European Patent Office Proposed by the Consultative Assembly of the Council of Europe. SG/D (50) 3 (Strasbourg, July 20, 1950); Committee of Experts on Patents—Analytical Summary of the Principal observations made by governments or experts on the draft proposal for a Convention for the creation of a European Patent Office. CM/WP IV (51) 1 (Strasbourg, December 7, 1950).

[8] European Convention relating to the Formalities required for Patent Applications (Paris, December 11, 1953) ETS No. 16; European Convention on the International Classification of Patents for Inventions (Paris, December 19, 1954) ETS No. 17.

[9] *Convention on the Unification of Certain Points of Substantive Law on Patents for Invention*, opened for signature 27 November 1963, (entered in force 1 August 1980) [Strasbourg Convention].

[10] Ibid.

[11] *European Patent Convention*, opened for signature 5 October 1973, (entered in force 7 October 1977).

[12] Ibid.

[13] *Marrakesh Agreement Establishing the World Trade Organization*, opened for signature 15 April 1994, 1867 UNTS 3 (entered into force 1 January 1995) annex 1C (*Agreement on Trade-Related Aspects of Intellectual Property Rights*) ('*TRIPS*').

[14] Epo.org. 2018. *EPO—Validation of European patents in Cambodia (KH) with effect from 1 March 2018*. [online] Available at: https://www.epo.org/law-practice/legal-texts/official-journal/information-epo/archive/20180209.html [Accessed 8 May 2022].

[15] *European Patent Convention*, opened for signature 5 October 1973, (entered in force 7 October 1977).

[16] Ingeborg Spillmann-Fürst, 'Europe edges closer to GMO patent harmonization' (1999) 17(9) *Nature Biotechnology* 842–843.

EU country can file a regional patent application through country-specific patent offices.[17]

5.2.1 How Do the EPC and EU Directives Operate?

Similar to the patent systems of other jurisdictions around the world, such as Australia and the US, the European patent system has been evolving (in terms of judicial decisions) in response to technological advancement. However, the emergence of innovative technologies, particularly related to biological inventions with commercial biotechnology applications, has brought a slew of complicated and contentious issues into the public sphere, perhaps emphasising the need for an update in the European patent system.[18] In contrast to the Australian and US patent systems, the European patent system includes different legislative instruments which also include a range of EU directives and regulations. The EPC is a regional patent convention and has an unusual position of overlapping with substantive EU law through the Biotechnology Directive. The EPC is not a part of a wider body of non-patent law, apart from other international agreements such as TRIPS.[19]

The EU directives and regulations operate under a distinct framework compared to the EPC, as they fall under the administration of the EU's European Parliament and European Council and are designed to harmonise laws across EU member states. For instance, the Biotechnology Directive is part of a broader body of EU law that goes beyond patent regulations and must align with fundamental EU principles, subject to interpretation by the European Court of Justice (ECJ) based on various principles enshrined in the EU treaties. While EU directives are initially formulated for EU members, their impact can extend beyond the EU if non-EU members also adopt policies influenced by EU directives. For example, Switzerland and Norway are not members of the EU, and EU directives and regulations have no effect on them. But as they are members of the EPC (which complies with the EU directives),[20] and hence, they must comply with the provisions of the EPC.[21]

The EPC established the EPO as a public international organisation that provides a centralised patent prosecution process for applicants who are interested in getting

[17] Richard Spencer, 'The German Patent Office' (1949) 31 *Journal of the Patent and Trademark Office Society* 79.

[18] Evisa Kica and Nico Groenendijk, 'The European patent system: dealing with emerging technologies' (2011) 24(1–2) *Innovation: The European journal of social science research* 85–105; Les Levidow et al., 'Regulating agricultural biotechnology in Europe: harmonisation difficulties, opportunities, dilemmas' (1996) 23(3) *Science and Public Policy* 135–157.

[19] *Marrakesh Agreement Establishing the World Trade Organization*, opened for signature 15 April 1994, 1867 UNTS 3 (entered into force 1 January 1995) annex 1C (*Agreement on Trade-Related Aspects of Intellectual Property Rights*) ('*TRIPS*').

[20] *European Patent Convention*, opened for signature 5 October 1973, (entered in force 7 October 1977).

[21] Ibid.

patent protection in all member states.[22] Additionally, patents registered with the EPO have a national effect across all signatory states.[23] The EPO operates Boards of Appeals at various levels to determine on appeals on decisions of patent examiners. Decisions made by the receiving section, examining divisions, opposition divisions, and legal division of the EPO are subject to independent review by the Boards of Appeals.[24] The Technical Board of Appeals (TBA) is responsible for hearing decisions of the examining divisions rejecting European patent applications or granting European patents, as well as decisions of the opposition divisions.[25] Referrals may be made to the Enlarged Board of Appeals (EBA) in order to assure consistent application of the provisions of EPC or in the event that a significant legal issue arises.[26] Specifically, the EPO President may refer a matter if it involves ensuring the consistent application of the EPC or an important legal issue, or when two Boards of Appeal have reached conflicting conclusions.[27] When the EBA receives a referral from the EPO President and there are no open appeal proceedings, the EBA operates as an advisory body and provides an opinion rather than a decision on the relevant legal issue.[28]

5.3 Biotechnology Directive

Since the 1980s, the European Commission had sought to enact a Directive concerning the patentability of biotechnological inventions.[29] The Commission was compelled by the diverging protection across member states and the greater protections in jurisdictions such as US and Japan, for example.[30] In July 1998, the European Council and the European Parliament approved a Biotechnology Directive that introduced the concepts of biotechnology for the first time into the European patent system.[31] The Biotechnology Directive aimed to address the complexities surrounding the patenting of biotechnology-related subject matter in both EU and

---

[22] Thomas R. Nicolai, 'The European Patent Convention: A Theoretical and Practical Look at International Legislation' (Pt HeinOnline) (1971) 5 *International Lawyer* 135.

[23] Office, E., 2022. *European Patent Office*. [online] Epo.org. Available at: https://www.epo.org/ [Accessed 2 April 2022].

[24] *European Patent Convention*, opened for signature 5 October 1973, (entered in force 7 October 1977), Article 21.

[25] Ibid.

[26] Ibid. Articles 22, 112(1).

[27] Ibid. Article 112(1)(b).

[28] Ibid.

[29] Directive 98/44/EC of the European Parliament and of the Council, on the legal protection of biotechnological inventions, *Parliamentary Debates*, 06 July 1998.

[30] Jasemine Chambers, 'Patent Eligibility of Biotechnical Inventions in the United States, Europe, and Japan: How Much Patent Policy Is Public Policy' (2002) 34 *The George Washington International Law Review* 223.

[31] Directive (n 29) Article 13.

EPC nations. Its primary objective was to harmonise patent law across EU member states, specifically in the field of biotechnology, ensuring that biotechnology products are patentable throughout the EU when appropriate. The directive achieves this by establishing clear regulations for the patenting of biotechnological products and methods, including those containing human-derived substances, and defining the scope of protection offered by these patents.

The aim of the Biotechnology Directive was to set out the standards for protection of patents across biotechnology subject matter.[32] In March of 1995, the European Parliament rejected the proposed Directive despite a common position being adopted on it. The main reason was it removed the barriers to patenting life forms.[33] However, the Commission presented an amended proposal later that year that sought to consider the ethical dimensions of biotechnology subject matter patentability. The European Parliament tabled 66 amendments that were put to the Commission in August 1997, with the Directive passing on 30th July 1998.[34] Many of its Articles have been implemented into the Implementing Regulations of the EPC to ensure that the approach of the EPC to biotechnology inventions is in parallel with the Directive.[35] The Biotechnology Directive is the only area of substantive law where the EPC and EU law overlap. The Directive itself requires that member states protect biotechnology inventions under national patent laws.[36]

The Biotechnology Directive required enactment through national legislation by July 30, 2000, and as of 15 January 2007, all 27 EU member states had implemented the Biotechnology Directive.[37] In 2001, the Netherlands attempted to challenge the Biotechnology Directive on several grounds but failed.[38] The ECJ dismissed claims that the Biotechnology Directive's adoption violated European procedural standards or that Article 95 of the European Commission Treaty was the incorrect legal basis for the directive.[39] The ECJ emphasised that the harmonisation achieved through the directive was justified and could only be effectively attained through collective Community action, not through the individual actions of member states.[40] While offering limited flexibility, the language of the Biotechnology Directive also provided much-needed legal certainty. Notably, the directive placed a firm emphasis on protecting human dignity and integrity by expressly denying the patenting of

[32] Ibid.

[33] R. Stephen Crespi, 'The European Biotechnology Patent Directive is Dead'(1995) 13(5) *Trends in Biotechnology* 162–164.

[34] Arthur Rogers, 'European Union approves gene-patent Directive' (1998) 351(9114) *The Lancet* 1500.

[35] *European Patent Convention*, opened for signature 5 October 1973, (entered in force 7 October 1977).

[36] Directive (n 29) Article 1.

[37] 'State of Play of the Implementation of Directive 98/44/EC', *Ec.Europa.Eu* (Webpage, 2007) https://ec.europa.eu/docsroom/documents/14892/attachments/1/translations/en/renditions/pdf.

[38] *Kingdom of the Netherlands v European Parliament and Council of the European Union* (2001) C-3 77/98.

[39] Ibid., 2.

[40] Ibid., 54.

the human body.[41] The Biotechnology Directive exclusively addressed the grant of patents, not other legal restrictions or prohibitions applying to research into patentable products or the exploitation of patented products.[42] Although the Biotechnology Directive did not explicitly define a biotechnological invention, it is widely recognised to involve the creation of a new life form by modifying the genetic code of an existing organism.[43]

The Biotechnology Directive was introduced in response to the rapid progress of biotechnology, and it mandates regular reporting on specific biotechnology and genetic engineering issues. The Commission established an expert group with the purpose of providing expertise, analysis, and opinions on current scientific breakthroughs and their implications on patent law.[44] The primary goal of the Biotechnology Directive was to foster harmonisation within the European patent law framework and establish clear and robust standards for safeguarding biotechnological innovations. However, in practice, national implementations have sometimes led to divergence, particularly in cases involving gene sequencing.[45] The Biotechnology Directive is narrower in scope than the title suggests, as it is concerned only with 'biological material.' Many of the intended benefits of harmonisation existed by virtue of EPO decisions by the time the Biotechnology Directive was enacted. To further align with the directive's objectives, the EPO introduced new rules (Rules 26 to 29 in Part II Chapter V of the EPC Implementing Regulations) that stipulate the directive should be used as a 'supplementary means of interpretation' alongside the EPC.[46]

The Biotechnology Directive incorporates many provisions that align with pre-existing national and EPO practices. Article 6(1) of the Directive, along with Recitals 36–39, 43, and 44, reflects the essence of Article 53(a) of the EPC. Additionally, Article 6(2) and Recitals 40–42 provide a non-exhaustive list of unpatentable subject matter based on ethical and moral considerations.[47] Although the Directive binds EU states but not the EPO itself, the EPO subsequently aligned its rules to accord with the ethical stances taken pursuant to the Directive. In following its decision in *Oncomouse*,[48] the EPO decided in *Leland Stanford/Modified Animal* that its rules should

[41] Ibid., 77.

[42] Ibid., 80.

[43] E. Richard Gold and Alain Gallochat, 'The European biotech directive: past as prologue' (2001) 7(3) *European Law Journal* 331–366.8).

[44] 'Commission Decision On Setting Up A Commission Expert Group On Development And Implications Of Patent Law In The Field Of Biotechnology And Genetic Engineering', *Ec.Europa.Eu* (Webpage, 2012) https://ec.europa.eu/docsroom/documents/14893/attachments/1/translations.

[45] Geertrui Van Overwalle, 'The implementation of the biotechnology directive in Belgium and its aftereffects: the introduction of a new research exemption and a compulsory license for public health' (2006) 37 *International Review of Intellectual Property and Competition Law (IIC)* 889–920.

[46] Part II Chapter V to the *EPC Implementing Regulations,* Preamble.

[47] Directive (n 29) Article 6(1), 6(2).

[48] *Harvard Oncomouse* T19/90 [1990] OJ EPO 476, 5; This case is discussed in Sect. 5.4.2 of this chapter in relation to ordre public or morality exception for patentability assessment.

conform to the new (at the time) Directive; therein allowing genetic modifications to an animal.[49] Additionally, the EPO held in *Harvard/Transgenic Animals* that even if an invention complied with an EPC rule [i.e., Article 6(2)(d)] it must also comply with Article 53(a) of the EPC.[50] In relation to the ethics debate, Recital 26 of the Biotechnology Directive touches on the issue of 'informed consent' with material that has a human origin; namely, the material (stem cells) that is used in developing bioprinted tissues and bioinks. In the case of *Breast and Ovarian Cancer v University of Utah*, the EPO rejected an argument that a failure to demonstrate consent made the subject matter unpatentable as per Article 53(a).[51] This is notably in terms of arguing for bioprinted tissue's validity although warrants further discussion in the order public or morality debate.[52]

5.3.1 Relevance to Bioprinted Tissues and Bioinks

The primary focus of this section is to evaluate the Biotechnology Directive's impact on the patentability of bioprinted tissues and bioinks. Notably, one of the principal aims of the Biotechnology Directive was to clarify the application of the EPC exceptions to biotechnology inventions. The Biotechnology Directive expressly states that new inventions which involve an inventive step and are capable of industrial application shall be patentable even if they comprise a product that consists of or contains biological material or a process in which biological material is processed, produced, or used.[53] Biological material, as defined in Article 2(1)(a), refers to material isolated from its natural environment or produced through technical processes and can be subject to an invention, even if it naturally existed before.[54] This is intended to clarify the distinction between a mere discovery of unpatentable biological subject matter and its industrial application. The EPC amended its Implementing Regulations to address any inconsistencies with the Biotechnology Directive, aligning its approach with the Board of Appeal's stance on subject matter patentability in biotechnology inventions.[55]

Article 5(1) of the Biotechnology Directive indicates that the human body at the various stages of its formation and development, and the simple discovery of one of its elements, including the sequence or partial sequence of a gene, cannot constitute patentable subject matter.[56] Article 5(2) of the Biotechnology Directive further provides that an element isolated from the human body or otherwise produced by a

[49] *Leland Stanford/Modified Animal* [2002] EPOR 16.

[50] *Harvard/Transgenic animal* T-315/03 [2005] EPOR 31.

[51] *Breast and Ovarian Cancer/University of Utah* T-1213/05 [2007] at 47.

[52] See Sect. 5.4.2 of this chapter.

[53] Directive (n 29) Article 3(1).

[54] Ibid. Article 2(1)(a).

[55] Gold and Gallochat (n 43).

[56] Directive (n 29) Article 5(1).

technical process may constitute a patentable invention even if its structure is that of a natural element.[57] For bioprinted tissues and bioinks, despite resembling the structure and composition of natural tissues, their production by technical processes may render them patentable. Considering the conclusions in Chap. 3 (considering the Australian perspective) of this book, a similar claim may satisfy the 'manner of manufacture' requirements, as the claimed bioprinted tissue and/or bioink is produced by a technical process and is considered an 'artificially created state of affairs.' On the other hand, Chap. 4's analysis (US perspective) indicates that if bioprinted tissues and/or bioinks are similar to natural tissues, they may not fulfil the 'markedly different characteristics' requirements and might not be patentable in the US.

Article 16(c) of the Directive requires the European Commission to provide reports on biotechnology patent law's impact and development. However, only two reports have been issued so far, focusing on the ethical use and patentability of human stem cells and cell lines.[58] The Commission stated in the 2005 reports that 'it is premature to give further definition or provide further harmonisation in this area' other than that provided by the Directive.[59] Separately, the European Group of Ethics opined in May 2002 that there was no ethical reason for a complete ban on patent applications with subject matter related to tissue engineering or stem cell lines.[60] However, the EPO Opposition Division's decision in the *Edinburgh* patent case limited stem cells to animal cells, disregarding the Ethics Group's stance.[61] The stem cells were considered patentable subject matter under Article 5(2) of the Biotechnology Directive as they were isolated elements produced by technical processes, not related to industrial or commercial use of embryos (Article 6(2)(c) of the Biotechnology Directive).[62]

The subject matter debate and the use of stem cells have evolved in case law in ways that are directly applicable to bioprinted tissues and bioinks. The EPO has held that stem cells such as induced pluripotent cells[63] and adult stem cells are patentable

---

[57] Ibid. Rule 29(2).

[58] *Report from the Commission to the European Parliament and the Council—Development and implications of patent law in the field of biotechnology and genetic engineering* (Commission of The European Communities No 52002DC0545, 07/10/2002) ('*Report from the Commission to the European Parliament and the Council—Development and implications of patent law in the field of biotechnology and genetic engineering*'); *Report from the Commission to the the Council and European Parliament—Development and implications of patent law in the field of biotechnology and genetic engineering* (Commission of The European Communities No 52005DC0312, 14/07/2005) ('*Report from the Commission to the the Council and European Parliament—Development and implications of patent law in the field of biotechnology and genetic engineering*').

[59] Ibid.

[60] *Opinion of The European Group on Ethics in Science and New Technologies to the European Commission—Ethical Aspects of Patenting Inventions Involving Human Stem Cells* (The European Group on Ethics in Science and New Technologies, 07 May 2002) ('*Opinion of The European Group on Ethics in Science and New Technologies to the European Commission—Ethical Aspects of Patenting Inventions Involving Human Stem Cells*').

[61] Opposition Division of the EPO (2003) Edinburgh (EP0695351B1).

[62] Directive (n 29) Articles 6(2)(c), 5(2).

[63] Induced pluripotent stem cells (also known as iPS cells or iPSCs) are a type of pluripotent stem cell that can be generated directly from a somatic cell; Qi Gu et al., 'D bioprinting human

subject matter, as are inventions for therapeutic and diagnostic purposes. In the TBA case of *Culturing Stem Cells/Technion* it was held that,

> Inventions which make use of publicly available human embryonic stem cell lines which were initially derived from a process resulting in the destruction of the human embryo are excluded [subject matter] under the provisions of Article 53(a) EPC in combination with Rule 28(c) EPC [...][64]

In essence, the central question revolved around whether cell lines were derived from processes that involved the destruction of human embryos. This scenario was inevitable before the advent of different cell lines and subsequent bioprinting of tissues and bioinks. While the EPO is not bound by ECJ decisions, it was likely influenced in the context of bioprinting-related subject matter. In *Oliver Brüstle v Greenpeace eV* (*Brüstle*), the subject matter was excluded from patentability where the technical teaching of the patent necessitates the destruction or use of human embryos.[65] The patent in the *Brüstle* case claimed a process that requires the destruction of a human embryo at the blastocyst stage in order to remove a stem cell.[66] The court provided a reasonable interpretation of an embryo, concluding that, 'any human ovum must, as soon as fertilised, be regarded as a human embryo ... if that fertilisation is such as to commence the process of development of a human being.'[67] Notably, current advancements in the field allow stem cell production without embryo destruction, thus rendering the subject matter beyond the scope of Article 6(2)(c) objections [of the Biotechnology Directive].[68]

Article 8 of the Biotechnology Directive establishes limitations to the scope of protection for patent applications related to biological material.[69] When biological material qualifies for patent protection under the Directive, the patent's scope is confined to material produced from that biological material through propagation or multiplication, retaining the same characteristics.[70] For example, if a bioink is the subject of a patent claim and that bioink is used to create another bioink with the same characteristics through propagation or multiplication, that patent application may also have a broader scope of protection. Similarly, if a patent application has claims directed towards a process of bioprinting a tissue or a process of formulating a bioink, such protection may extend to biological material directly obtained from that process and any subsequent generations possessing the same traits.[71] Further, if a bioprinted tissue is developed using a particular bioprinting process, patent protection may also extend to the bioprinted tissue developed using the patented bioprinting

induced pluripotent stem cell constructs for in situ cell proliferation and successive multilineage differentiation' (2017) 6(17) *Advanced healthcare materials* 1700175.

[64] *Culturing Stem Cells/Technion* T 2221/10 [2014].

[65] *Oliver Brüstle v Greenpeace eV* C-34/10 [2011].

[66] Ibid., 15.

[67] Ibid., 35.

[68] Directive (n 29) Article 6(2)(c).

[69] Ibid. Article 8.

[70] Ibid. Article 8(1).

[71] Ibid. Article 8(2).

process. This means that if a patent application only claims the bioprinting process and not the tissue itself, Article 8(2) of the Biotechnology Directive allows protection to extend to unclaimed bioprinted tissue derived from the patented process.

Article 9 of the Biotechnology Directive is also relevant to the present discussion. Article 9 details that,

> The protection conferred by a patent on a product containing or consisting of genetic information shall extend to all material, save as provided in Article 5(1), in which the product incorporated and in which the genetic information is contained and performs its function.[72]

The case of *Monsanto Technology v Cefetra* sheds light on the application of Article 9 of the Biotechnology Directive.[73] The plaintiff held a patent for enzymes isolated from bacteria and inserted into a soy plant, making it herbicide-resistant. However, when soy meal made from these plants was imported to the EU, the CJEU ruled that Article 9 did not protect it, as the enzymes couldn't fulfil their patented purpose in the dead material. This precedent highlights that Article 9 doesn't safeguard all genetically modified products but only those capable of providing the patented benefit.[74] Stem cells can be bioprinted in the form of a bioink in combination with other sensitive biological components such as genes and proteins to regulate cellular function, and if a patent application is directed towards such a gene or protein that regulates the function of cells incorporated in a bioink, patent protection may extend to the bioink and the corresponding bioprinted tissue or scaffold.[75] However, if such gene or protein for which the patent protection is conferred is unable to perform its function (regulate cellular function) in the bioink in which it is incorporated, the protection may not extend to the bioink or corresponding bioprinted tissue or scaffold in which it is incorporated.[76]

5.4 Patentable Subject Matter—Bioprinted Tissues and Bioinks Perspective

5.4.1 Patentable Inventions According to Article 52 of EPC

The patentable subject matter requirements play a crucial role in assessing EPC patent applications, acting as a primary examination criterion before considering other patentability aspects like novelty, inventive step, industrial application, disclosure, and support requirements.[77] Article 52(2) of the EPC is significant, as it precludes

[72] Ibid. Article 9.

[73] *Monsanto Technology v Cefetra* 428/08 [2010] ECR I-6765.

[74] Directive (n 29) Article 9.

[75] Weijie Peng et al., 'Challenges in bio-fabrication of organoid cultures' (2018) *Cell Biology and Translational Medicine, Volume 3* 53–71.

[76] *Monsanto Technology v Cefetra* 428/08 [2010] ECR I-6765.

[77] Ibid. Article 52(1).

inter alia discoveries, scientific theories, mathematical methods, aesthetic creations, schemes, rules and methods for performing mental acts, playing games or doing business, computer programs, and presentations of information from patenting 'as such.'[78] The EPO's definition of the term 'invention' has undergone a complex evolution, initially shaped by TBA decisions concerning Article 52(2).[79] In particular, the decisions made by TBA laid the foundation for all future developments by interpreting Article 52(2) as a definitional provision that resolves to a single requirement for technical character.[80] While the TBA case law from the 1980s and 1990s largely dealt with computer-related inventions and did not give a meaningful analogy to biological inventions, the rulings have clarified how 'as such' in Article 52(2) is to be interpreted. The wording is mainly focused on Article 52(2) as it restricts inventions claiming excluded subject matter 'as such.' The following discussion analyses the provisions of Article 52(2) of the EPC and how 'as such' has been interpreted by the TBA in different cases.

Viacom/Computer-related Invention was the first substantial TBA case to interpret the phrase.[81] The TBA decided that a computer-aided design (CAD) program was patentable since it was not 'as such' related to the mathematical method exclusion category.[82] By enhancing computer processing speed and practically modifying the computer, the CAD program demonstrated a 'technical effect.' Consequently, the invention was not deemed related to an Article 52(2) exclusion 'as such' due to its 'technical effect.'[83] The TBA further concluded that the invention was not related to an Article 52(2) exclusion 'as such' because it had a 'technical effect.'[84] In *IBM/Computer Product Program*, the TBA refined the 'technical effect' assessment.[85] In this case, the TBA clarified that software operating on a computer is not enough to be considered a 'technical effect' in and of itself.[86] Further, the TBA detailed that software that creates 'further technical effects' and 'solves a technical problem' could be patentable.[87] A computer programe product, in particular, 'is not excluded from patentability' if it generates a technological effect 'beyond the "normal" physical interactions between program (software) and computer (hardware).'[88] Software,

---

[78] Ibid. Articles 52(2), (3).

[79] The EPO Technical Board of Appeals interprets EPC statutes. The decisions of this judicial body determine the substantive EPO patent examination process. In contrast, rulings of individual EPC member state national courts do not need to bind to the Technical Board of Appeals or the EPO patent examination system. Furthermore, while each member state adheres by the EPC's requirements, they are not bound by decisions made by other member states or by the EPO Technical Board of Appeals.

[80] *Christian Franceries/Traffic Regulation* ('Christian Franceries') T16/83 [1988] EPOR 65.

[81] *Viacom/Computer related invention* T 0208/84 [1986] EPOR 74.

[82] Ibid.

[83] Ibid.

[84] Ibid.

[85] *IBM/Computer Product Program* T 1173/97 [1998] EPOR 219.

[86] Ibid.

[87] Ibid.

[88] Ibid.

for example, is patentable where 'a technical effect … is achieved by the internal functioning of a computer itself under the influence of said program,' according to the EPC.

The case analysis illustrates the application of the 'as such' doctrine and the 'technical effect' approach in determining the patentability of computer-implemented inventions. These principles are also relevant for 3D bioprinting, which employs CAD software to design and print bio-functional materials in a layer-by-layer fashion to form complex 3D tissues and organs. The term 'bioink,' which denotes the combination of living cells, biomaterials, or active biomolecules used for 3D bioprinting, is not excluded from patentability 'as such' because it contributes to a technical effect of creating tissue-like 3D structures. Moreover, the bioink should also meet the criteria of solving a technical problem and producing further technical effects, such as replicating natural tissue properties and supporting cell functions.

Additionally, the TBA's interpretation of the Article 52(2) exclusion and the term 'as such' holds relevance for our analysis, particularly concerning subject matter related to bioprinted tissues and bioinks. As one of the major essential elements of bioprinted tissues and bioinks is stem cells, subject matter directed towards bioprinted tissues and bioinks may be excluded as discoveries in light of Article 52(2) if the claims are directed towards stem cells 'as such.'[89] However, the specific subject matter may not be excluded under Article 52(2) if a claim directed towards bioprinted tissues and bioinks *includes* stem cells and the specific claim as a whole has a practical outcome (such as bioprinted tissue or bioink), resulting in a 'technical effect.' For example, the 'technical effect' of bioprinted tissues and bioinks may manifest in their medical applications, such as tissue replacement, wound healing, or pharmaceutical testing.[90] Another argument is that subject matter related to bioprinted tissues and bioinks are resulted from a technical, man-made process using bioprinters or similar systems, satisfying the technicality criteria of Article 52(2). For example, a patent application with claims directed towards a bioprinted liver tissue may meet the Article 52(2) technicality criteria, if specific bioprinted liver tissue is claimed in a way reflecting that it is developed using a bioprinter and hence it is an outcome of a technical, man-made process.[91]

In 2002, a pivotal case, *Relaxin/Howard Florey Institute*, came before the TBA, which involved claims of a patent application titled 'Molecular cloning and characterization of a further gene sequence coding for human relaxin.'[92] In this case, the appellants raised patentability arguments on different grounds, particularly based on

[89] *European Patent Convention*, opened for signature 5 October 1973, (entered in force 7 October 1977), Article 52(2); See Sect. 5.3.1 of this chapter.

[90] Mario Chopin-Doroteo, Edna Ayerim Mandujano-Tinoco and Edgar Krötzsch, 'Tailoring of the rheological properties of bioinks to improve bioprinting and bioassembly for tissue replacement' (2021) 1865(2) *Biochimica et Biophysica Acta (BBA)-General Subjects* 129782; Yuzhen Wang et al., 'Tailoring bioinks of extrusion-based bioprinting for cutaneous wound healing (2022) *Bioactive Materials*.

[91] Hideki Kizawa et al., 'Scaffold-free 3D bio-printed human liver tissue stably maintains metabolic functions useful for drug discovery'(2017) 10 *Biochemistry and biophysics reports* 186–191.

[92] *Relaxin/Howard Florey Institute* T 0272/95 [2002] EPOR.

Article 52(2)(a), discoveries not being patentable subject matter.[93] The appellants contended that the invention's focus was on identifying the genetic sequence of the H2-relaxin gene, and the proprietor had essentially 'cracked the code' by obtaining a code book from genetic material donors.[94] The appellants further added that the subject matter claimed was nothing more than a discovery of the properties of a substance that had most likely been in nature for thousands of years. According to Article 52(2)(a) of the EPC, the appellants asserted that such an invention based on discovery was not eligible for patent protection.[95]

In response to the patentability arguments, the respondents/patentees put forth a compelling interpretation based on Rule 23(e)(2) EPC (now Rule 29), contending that patent protection should indeed encompass elements isolated from the human body or produced through a technical process, even if their structure mirrors that of a natural element. Relying on this provision, they argued that the claimed H2-relaxin DNA should be considered patentable.[96] The TBA carefully analysed Article 52(2)(a) in conjunction with implementing Rules 23(d) and 23(e)(2) EPC, which explicitly stated that elements isolated from the human body or produced using technical means, including gene sequences, may constitute patentable inventions, even if their structure is identical to that of a natural element.[97] The TBA concluded that, 'It follows from the text itself that the matter mentioned above is not to be considered as an exception to patentability under Article 53(a) EPC. Claims 2 to 7, 10 to 14 and 18 to 21 are, thus, allowable under this article.'[98]

The issue of patentability of biological material was again analysed by the TBA in 2007 in *Breast and ovarian cancer/university of Utah*, where claims of a patent application entitled '17q-Linked breast and ovarian cancer susceptibility gene' were appealed on the basis of Article 52(2) of EPC.[99] This European patent application belonged to the same patent families as those at issue in *AMP* in the US and *D'Arcy* in Australia. The major argument by the appellant was that the sequences of the probes in claim one occurred in nature and were thus a discovery rather than an invention. Hence, they argued that the claimed subject matter was not patentable under Article 52(2)(a) EPC. Considering these arguments, the TBA cited the decision of *Relaxin/ Howard Florey Institute* and indicated that Article 52(2)(a) EPC is to be interpreted in accordance with implementing Rule 23e (2) EPC. In light of this, the TBA again made it clear that claims 1 to 3 are about nucleic acid probes that contain partial DNA sequences of the human BRCA1 gene that were obtained by technical processes. As a result, these probes constitute isolated human body elements, which are patentable

[93] *European Patent Convention*, opened for signature 5 October 1973, (entered in force 7 October 1977), Article 52(2)(a).

[94] Ibid., XII.

[95] Ibid.

[96] Ibid., XIII.

[97] Ibid., 7.

[98] Ibid.

[99] *Breast and ovarian cancer/university of Utah* T 1213/05 [2007] EPOR.

subject matter. As a result, the subject matter of claims 1 to 3 did not fall under the class of inventions that are not patentable as discoveries (Article 52(2)(a) EPC).[100]

The TBA's decisions in *Relaxin/Howard Florey Institute* and *Breast and ovarian cancer/university of Utah* are relevant to the current discussion on Article 52(2) of the EPC and its relevance to the patentability of subject matter related to bioprinted tissues and bioinks. Stem cells and other elements from the human body are essential components of both bioprinted tissues and bioinks. Should a claim encompassing bioprinted tissues or bioinks feature such elements, isolated from the human body or produced through a bioprinting process, it could potentially meet the criteria of Article 52(2) and thus be deemed patentable. Further, the TBA's statement that '... even if the structure of that element is identical to that of a natural element' holds particular significance in our discussion. Understanding the precise scope of the term 'element' becomes crucial. For example, if a claim pertains to a bioprinted tissue that closely resembles and functions like natural tissue, how would it be assessed in light of the EPC provisions? In this scenario, the bioprinting method would qualify as a 'technical process,' and the resulting tissue may be deemed patentable, even if its structure replicates that of natural tissue.

The EPO's stance on the patentability of elements isolated from the human body, including genetic material, diverges significantly from the conclusions reached by Australian and US courts (as explored in Chaps. 3 and 4 of this book). In Australia, the courts have determined that claims directed towards isolated nucleic acids fail the manner of manufacture requirement and are inherently non-patentable under s 18(1)(a) of the *Patents Act*.[101] The patentability of such claims in Australia is assessed through two filters: identifying the substance of the claim and considering whether that substance has been made or changed by human action or is artificial.

On the other hand, the US Supreme Court held that genes and the information they encode are not patent eligible under Section 101 simply because they have been isolated from the surrounding genetic material.[102] However, the US Supreme Court in the same decision has also considered cDNA (complementary DNA) as patentable, as it is considered to be an 'artificial creation.' These distinctions demonstrate that patent applications featuring claims directed towards elements isolated from the human body 'as such' would be considered patentable by the EPO, but not in Australia and the US. Considering the relevant case decisions discussed above, it appears that claims directed towards bioprinted tissues may meet the requirements of Article 52 of the EPC and, thus, may be deemed patentable by the EPO. This divergence in approaches highlights the complexity and nuances in assessing the patentability of bioprinted tissues and underscores the need for careful consideration and harmonisation of patent laws in line with the emerging technologies.

[100] Ibid., 45.

[101] See Sect. 3.1.1 of Chap. 3.

[102] See Sect. 4.3.4 of Chap. 4.

5.4.2 *The* Ordre Public *or Morality Exception*

5.4.2.1 Article 53(a) of EPC

Under Article 53(a) of the EPC, inventions whose exploitation is not in accordance with the generally accepted standards of conduct are to be excluded from patentability as being immoral. The idea of morality refers to the belief that certain behaviours are proper and acceptable, while others are immoral, and that this attitude is based on the entirety of accepted norms that are deeply ingrained in a culture.[103] The culture in question is the culture inherent in European society and civilisation for the purposes of the EPC. In Europe, there are two types of morality provisions. The first is found in the EPC's Article 53(a) and the second is in the Biotechnology Directive's Article 6(1).[104] Specifically, Article 6 of the Biotechnology Directive guides the interpretation of Article 53 of the EPC for inventions related to biotechnology. Article 53 of the EPC sets forth certain exclusions from patentability, notably barring patents for inventions contrary to morality. Specifically, Article 53(a) stipulates that EU patents shall not be granted for inventions whose commercial exploitation would run counter to *ordre public* or morality. Importantly, such contravention cannot be solely based on the fact that it is prohibited by law or regulation in some or all of the Contracting States. In the context of recent biotechnological advancements, like bioprinting, the significance of Article 53(a) has amplified, leading to debates on its efficacy in addressing moral issues.[105] Consequently, the provisions of Article 53 of the EPC hold relevance for patent applications featuring claims directed towards bioprinted tissues and bioinks.

The EU's case law serves as a crucial point of reference for addressing the ongoing debate surrounding patentable subject matter requirements, particularly the *ordre public* considerations for biological inventions. Notably, in *Plant Genetic Systems*, the TBA held that the concept of *ordre public* also covers protecting our environment.[106] In this case, the TBA examined an invention pertaining to genetic engineering of plants and seeds to confer herbicide resistance. The argument raised was whether this subject matter could be patented under Article 53(a) and (b) of the EPC. The TBA held that advanced scientific inventions that intersect with morality extend beyond individuals and also encompass environmental considerations. If the commercial exploitation of an invention infringes upon social order or compromises the well-being of the environment, it should be excluded from patentability. In addition, it was held the concept of morality is tied to our understanding of what is right or

[103] Joseph Straus, 'Ordre public and morality issues in patent eligibility', *Intellectual property in common law and civil law* (Edward Elgar Publishing, 2013).

[104] *European Patent Convention*, opened for signature 5 October 1973, (entered in force 7 October 1977), Article 53(a); Directive 98/44/EC of the European Parliament and of the Council, on the legal protection of biotechnological inventions, *Parliamentary Debates*, 06 July 1998 (n Article 6(1).

[105] Jamil Ammar, 'The Medical Mile Gearing toward 3D-Bespoke Healthcare: A Comparison of United States and European Union Patent Regimes' (2016) 52 *Gonzaga Law Reveiw* 279.

[106] *Plant Genetic Systems N.V., et al. v Greenpeace Ltd.* T356/93 [1995] OJ EPO 545 at 5, 6.

wrong, that is, the norms rooted in our societies. In line with Article 53(a), inventions that violate these conventional standards of conduct in society are to be excluded from patentability. This case highlights the importance of understanding the broader impact of biotechnological innovations on both individuals and the environment. It underscores the need for a comprehensive framework that evaluates patentability not only from a technical standpoint but also from ethical and societal perspectives. Such considerations are crucial in shaping the future direction of biotechnology patents within the EU.

The second part of Article 53(a) contains the qualification that exploitation shall not be deemed contrary to *ordre public* merely if prohibited by law or regulation in some or all contracting states. This makes it clear that an assessment of whether a particular subject matter is to be considered contrary to *ordre public* is not dependent on national law or regulation. On the other hand, the TBA in *Plant Genetic Systems* held that it shall not be automatically regarded as complying with Article 53(a) of the EPC merely because it is allowed.[107] On the facts, the TBA held that there was no evidence of actual disadvantage, and the assessment of patentability regarding Article 53(a) should not be decided on the basis of 'balancing-exercises.'[108] A question, therefore, is whether the subject matter violates social norms. It is important to analyse whether bioprinted tissues and bioinks fall within this realm.

The EPO's examination of *ordre public* under Article 53(a) was notably illustrated in the groundbreaking *Oncomouse* case, predating the implementation of the Biotechnology Directive by almost eight years.[109] The case of *Oncomouse* (also known as Harvard Mouse) concerned one of the world's first transgenic animals. An animal that has had its genome altered by the transfer of a gene or genes from a different species or breed is said to be transgenic.[110] *Oncomouse* was genetically modified so it could develop cancer readily for the purposes of research. Harvard sought a patent for the *Oncomouse*, sparking ethical discussions in the realm of biotechnology and related fields of modern research. While Harvard was granted a patent for a 'transgenic non-human mammal,' the EPO carefully considered the prerequisites, particularly the non-human aspect.[111] In its assessment, the EPO evaluated both Article 53(a) and Article 53(b), which prohibits patents for 'animal varieties or essentially biological processes' for their production. Regarding Article 53(b), the EPO concluded that the *Oncomouse* did not fall within its scope, as it was not considered an animal variety. In addressing *ordre public*, the EPO adopted a utilitarian approach. In effect, the EPC asked whether the suffering of the mice was outweighed by the potential benefits to humankind.[112] These matters had previously come before the EPO in the *Upjohn Mouse* case, where a patent claimed a transgenic mouse that became bald for the

---

[107] Ibid. at 7.

[108] Ibid. at 18.8.

[109] *Harvard Oncomouse* T19/90 [1990] OJ EPO 476.

[110] Ibid.

[111] *Method for producing transgenic animals*, Priv L No EU, ch Harvard College, EP0169672B1 (1992) ('*Method for producing transgenic animals*').

[112] *Harvard Oncomouse* T19/90 [1990] OJ EPO 476, 5.

purpose of testing baldness and wool products.[113] However, the EPO weighed up the benefit in curing baldness against the pain suffered by the mice and found this to be contrary to morality. These cases demonstrate the intricacies involved in assessing patentability in biotechnology, where ethical implications play a significant role in the decision-making process.

The *ordre public* debate reared its head again in another case connected with the *Oncomouse*, in opposition proceedings initiated when the patentees amended the patent.[114] Of the numerous grounds argued, the most fundamental was directed to the application of Article 53(a) and its elaboration in the Administrative Regulations.[115] Rule 23(d) of the EPC incorporates key elements of the Biotechnology Directive, and notably, Rule 23(d)(d) specifies that inventions involving the genetic modification of animals causing them suffering are excluded from patentability. This provision aligns with Article 53(a) of the EPC, which addresses the concept of *ordre public*.[116] In its deliberations, the TBA ventured to determine whether the claimed subject matter would be contrary to *ordre public*, and in doing so, it explored the delicate balance between animal suffering and potential benefits to humankind. The TBA presented several propositions on this matter, seeking to establish a clear framework for assessing the moral implications as prescribed by Article 53(a). However, case law did not provide a definitive answer as to the prevailing moral attitude pursuant to Article 53(a). The TBA referred to its 1990 finding and stated that the real starting point for assessing the Article is to weigh suffering against mankind's benefit.[117] Throughout the process, the TBA engaged in a comprehensive examination of evidence and arguments and even sought policy and morality opinions to guide their decision-making. Ultimately, the TBA leaned towards favouring the benefits to mankind as a guiding principle for their assessment under Article 53(a).[118]

Upon closer examination of the implications of the *Oncomouse* case on patent applications concerning bioprinted tissues and bioinks, an intriguing avenue for exploration emerges—the interpretation of 'suffering versus mankind's benefit.' This concept might hold relevance in the current discussion. For example, if embryonic stem cells are used for producing bioprinted tissues and bioinks, and the embryonic stem cells are derived from an embryo, would this be considered to give rise to the suffering of an embryo? An embryo is the first stage of a multicellular organism's development. Embryonic development is the stage of the life cycle that begins shortly after conception and continues through the production of body structures such as tissues and organs in sexually reproducing organisms. Specifically, if the extraction

[113] *The Upjohn Company v Akzo Nobel Pharma B.V.* T791/96 [1999].

[114] *Harvard/Transgenic animal* T-315/03 [2005] EPOR 31.

[115] Implementing Regulations to the Convention on the Grant of European Patents of 5 October 1973 as adopted by decision of the Administrative Council of the European Patent Organisation of 7 December 2006 and as last amended by decision of the Administrative Council of the European Patent Organisation of 15 December 2020.

[116] *European Patent Convention*, opened for signature 5 October 1973, (entered in force 7 October 1977), Rule 23d(d).

[117] *Harvard Oncomouse* T19/90 [1990] OJ EPO 476.

[118] *Harvard/Transgenic animal* T-315/03 [2005] EPOR 31.

of embryonic stem cells from an embryo for producing bioprinted tissues and bioinks is considered to be suffering, how would the benefit of mankind (from bioprinted tissues and bioinks) be quantified against that suffering of the embryo? Further, how would it impact the patentability of specific bioprinted tissues and bioinks? To grasp the intricate relationship between embryonic stem cells, *ordre public*, and morality, the subsequent section's analysis aims to shed light on the patentability of these stem cells from a bioprinting perspective. Moreover, it endeavours to address the pertinent questions arising from this ethical conundrum.

5.4.2.2 Impact on Patentability of Embryonic Stem Cells

Embryonic stem cells are located in the inner cell mass of the human blastocyst, a stage of the developing embryo that lasts from the fourth to the seventh day after fertilisation. After this period, the embryonic stem cells naturally dissipate, giving way to the formation of the three primary tissue layers that lay the foundation for the body's intricate structures, including tissues and organs.[119] When scientists extract the inner cell mass and grow these cells under particular laboratory conditions, they retain embryonic stem cell characteristics.[120] Unlike embryonic stem cells, tissue-specific stem cells, also known as somatic or adult stem cells, are profoundly specialised, mainly generating diverse cell types within the specific tissue or organ they inhabit.[121]

The patentability of inventions with subject matter related to human embryonic stem cells is still a point of contention, not only because of the complex ethical considerations involved but also because such inventions may fall under the specific exclusions laid in Article 6(2)(c) of the Biotechnology Directive.[122] This is especially relevant for embryonic stem cells because they are produced by the destruction of an embryo (which may be seen as a prospective human with a derived right to life). As a result, morality and *ordre public*-based ethical concerns may still persist during patentability. While Article 6 sought to provide a comprehensive list of morally unpatentable subject matter at the time of its inception, the discord between national and European courts has sowed seeds of legal ambiguity, casting shadows over patenting strategies related to embryonic stem cells. As such, navigating this intricate landscape of ethics and patent law becomes an arduous task, where the course of action is far from definitive.[123]

The *ordre public* challenges that surround biotechnology are not new to the EPO, as several cases illustrate. Among them, the *Edinburgh Patent* decision stands

---

[119] Graziano Martello and Austin Smith, 'The nature of embryonic stem cells' (2014) 30 *Annual review of cell and developmental biology* 647–675.

[120] Ibid.

[121] Michaela Prochazkova et al., 'Embryonic versus adult stem cells', *Stem cell biology and tissue engineering in dental sciences* (Elsevier, 2015) 249–262.

[122] Directive (n 29) Article 6(2)(c); See Sect. 5.3.1 of this chapter.

[123] Aurora Plomer, Kenneth S. Taymor and Christopher Thomas Scott, 'Challenges to human embryonic stem cell patents' (2008) 2(1) *Cell Stem Cell* 13–17.

out, stemming from a patent application filed in 1994, which sought to protect the 'isolation, selection, and propagation of animal transgenic stem cells.'[124] A major public debate on the patenting of stem cell technology was triggered by the grant of this patent.[125] The opposition procedure brought against this patent invoked Article 53(a) in conjunction with Rule 23d(c) of the EPC, which explicitly excludes the patentability of human embryos used for industrial and commercial purposes. The opposition division, comprising three patent examiners and a legal expert, critically reevaluated the patent claims concerning the isolation and enrichment of animal stem cells, particularly those other than embryonic stem cells. They also scrutinised the claims relating to animal cells that could form a mix of various cells, including desired stem cells.[126] The Opposition Division determined that the issued patent did not meet the requirements of EPC Article 53(a) and Rule 23d(c). As a result, the patent was amended by the patentee, by removing the animal and human stem cells (other than embryonic).[127] Consequently, this decision had significant implications for embryonic stem cell research, restricting their patentability.

However, it is important to note that the restriction applied solely to embryonic stem cells, leaving room for inventions with claims directed towards other types of stem cells, such as adult stem cells (e.g., induced pluripotent stem cells), which possess different characteristics and may still be considered patentable. This decision confirms and clearly demonstrates that the exploitation of human embryos for industrial or commercial purposes (in this case, to produce human embryonic stem cells) is not patentable, as was made clear in one of the amendments made to the EPC as a result of the Biotechnology Directive in 2002.[128] While the selection of stem cells for bioprinting is highly application dependent, with many cell types being required, stem cells such as mesenchymal stem cells have been used for different types of tissue engineering and bioprinting applications.[129]

Human embryonic stem cells were also contested in the *Wisconsin Alumni Research Foundation* (*WARF*) case.[130] In *WARF*, The Examining Division rejected the applicant's patent application based on the provisions of Article 53 of the EPC.

---

124 *Isolation, selection and propagation of animal transgenic stem cells*, Priv L No EU, ch University of Edinburgh, EP0695351B1 (1999) ('*Isolation, selection and propagation of animal transgenic stem cells*').

125 Aurora Plomer et al., 'Stem cell patents: European patent law and ethics report' (2006) *FP6 Life Sciences, Genomics and Biotechnology for Health', SSA LSSB-CT-2004-005251. Nottingham: School of Law.*

126 Ibid. claims 1, 31.

127 EPO decides to maintain 'Edinburgh' patent in an amended form that no longer includes human or animal embryonic stem cells.

128 Directive (n 29).

129 Mesenchymal stem cells are multipotent adult stem cells found in a variety of organs such as the umbilical cord, bone marrow, and adipose tissue. Mesenchymal stem cells can divide and develop into a variety of tissues, including bone, cartilage, muscle, and fat cells, as well as connective tissue; Sindhuja D. Eswaramoorthy et al., '3D bioprinting of mesenchymal stem cells and endothelial cells in an alginate-gelatin-based bioink' (2021) 5(1) *Journal of 3D printing in medicine* 23–36.

130 *Wisconsin Alumni Research Foundation* T1374/04 [2006].

At the heart of the matter was the claim of a cell culture comprising stem cells derived from the destruction of an embryo[131] in the pre-implantation stage.[132] The patent application's first claim described a cell culture containing primate embryonic stem cells that exhibited specific characteristics, including the ability to proliferate in vitro for over a year, maintaining an unaltered karyotype representative of the primate species, preserving the potential to differentiate into various tissues, and being inhibited from differentiating when cultured on a fibroblast feeded layer.[133]

The application was denied because claims 1 to 7 and 9 and 10 did not fulfil the requirements of Article 53(a) EPC 1973 in connection with Rule 23d(c) [now 28(c)]. Considering this, the EPO's TBA referred the following questions of law to the EPO's EBA for decision:

> (1) Does Rule 23d(c) EPC apply to an application filed before the entry into force of the rule?
>
> (2) If the answer to question 1 is yes, does Rule 23d(c) EPC forbid the patenting of claims directed to products (here: human embryonic stem cell cultures) which – as described in the application – at the filing date could be prepared exclusively by a method which necessarily involved the destruction of the human embryos from which the said products are derived, if the said method is not part of the claims?
>
> (3) If the answer to question 1 or 2 is no, does Article 53(a) EPC forbid patenting such claims?
>
> (4) In the context of questions 2 and 3, is it of relevance that after the filing date the same products could be obtained without having to recur to a method necessarily involving the destruction of human embryos (here: e.g., derivation from available human embryonic cell lines)?[134]

The EBA set out its decision under several broad headings and, regarding question 1, indicated that, 'Rule 28(c) (formerly Rule 23d(c)) EPC applies to all pending applications, including those filed before the Rule came into force.' Adding to the above, the EBA further answered question 2 detailing that:

> Rule 28(c) (formerly Rule 23d(c)) EPC prohibits the patenting of claims to products which - as described in the application - at the time of filing could only be produced by a process which inevitably involves the destruction of human embryos from which the products are obtained, even if that process does not form part of the claims.[135]

---

[131] A multicellular organism's embryo is the first stage of development. Embryonic development is the stage of the life cycle that begins shortly after fertilisation and continues through the development of body parts such as tissues and organs in sexually reproducing animals. Each embryo begins life as a zygote, a single cell developed by the union of gametes (i.e., the process of fertilisation which is the fusion of a female egg cell and a male sperm cell). A single-celled zygote undergoes multiple fast cell divisions, known as cleavage, in the early stages of embryonic development to create a blastula, which resembles a ball of cells; Martin F. Pera, Benjamin Reubinoff and Alan Trounson, 'Human embryonic stem cells' (2000) 113(1) *Journal of cell science* 5–10.

[132] *Primate embryonic stem cells*, Priv L No EU, ch Wisconsin Alumni Research Foundation, EP0770125A4 (1997) ('*Primate embryonic stem cells*').

[133] Ibid. Claim 1.

[134] *Wisconsin Alumni Research Foundation* T1374/04 [2006].

[135] *Wisconsin Alumni Research Foundation* G0002/06 [2008].

Considering question 3, the EBA clarified that answering this question was unnecessary since questions 1 and 2 had already been addressed affirmatively. As for question 4, the EBA emphasised that the potential to manufacture the same products without destroying human embryos after the filing date did not impact the analysis.[136] The focus remained on whether the patent claimed subject matter related to human embryonic stem cells derived through embryo destruction, which would be contrary to the prohibition of commercial exploitation of human embryos. The patentee contended that the subject matter was the cell lines themselves, attempting to distance it from the destructive process. However, the EBA took into account the spirit of the invention, examining the process by which the cell culture was created. As the process necessitated the exploitation of an embryo, it fell under the exclusion provisions of Article 53(a) of the EPC.

These decisions have made it clear that EU patent applications with subject matter claiming human embryonic stem cells that are derived by a process resulting in the destruction of the embryos are excluded from patentability under the provisions of Article 53(a) of EPC in combination with Rule 28(c) EPC. However, an invention with subject matter related to human embryonic stem cells that have taken place at an earlier point and obtained without destroying human embryos may not fall within the scope of Article 53(a) and Rule 28(c). This distinction is highly significant, given that stem cells, particularly mesenchymal and embryonic stem cells, are key components of bioprinted tissues and bioinks.[137] For example, envision a team of scientists who have bioprinted a 3D human cardiac scaffold using cardiomyocytes derived from human embryonic stem cells, which are bioprinted into gelatin hydrogel on glass to assess viability and alignment.[138] On the basis of the EBA's analysis in the *WARF* case, such an invention may be excluded from patentability if the human embryonic stem cells that are used to derive cardiomyocytes are derived through the destruction of human embryos. In light of the court's conclusions in the *Brustle* case as discussed in Sect. 5.4.1 of this chapter, this subject matter may also fall within the scope of Article 6(2)(c) of the Biotechnology Directive and be considered unpatentable.[139] The implications of these decisions underscore the ethical complexities surrounding stem cell research, especially when it comes to bioprinted tissues and bioinks. The intersection of patent law and biotechnology ethics continues to be a subject of extensive debate, shaping the future landscape of patent protection in this rapidly advancing field.

---

136 Ibid.

137 Feng Xu et al., 'Embryonic stem cell bioprinting for uniform and controlled size embryoid body formation' (2011) 5(2) *Biomicrofluidics* 022207; Liliang Ouyang et al., 'Three-dimensional bioprinting of embryonic stem cells directs highly uniform embryoid body formation' (2015) 7(4) *Biofabrication* 044101; Mano Govindharaj et al., '3D Bioprinting of human Mesenchymal Stem Cells in a novel tunic decellularized ECM bioink for Cartilage Tissue Engineering' (2022) *Materialia* 101457.

138 Justin Liu et al., 'Rapid 3D bioprinting of in vitro cardiac tissue models using human embryonic stem cell-derived cardiomyocytes' (2019) 13 *Bioprinting* e00040.

139 Directive (n 29) Article 6(2)(c).

In 2014, the exclusion based on the use of embryonic stem cells was considered by the Court of Justice of the European Union (CJEU) in *International Stem Cell Corporation v Comptroller General of Patents, Designs and Trade Marks* case.[140] In this case the CJEU analysed two patent applications filed at the United Kingdom IP Office (UKIPO) that related to 'Parthenogenetic activation of oocytes for the production of human embryonic stem cells, claiming methods of producing pluripotent human stem cell lines from parthenogenetically activated oocytes and stem cell lines produced according to the claimed methods' and 'Synthetic cornea from retinal stem cells, claiming methods of producing synthetic cornea or corneal tissue, which involve the isolation of pluripotent stem cells from parthenogenetically activated oocytes, and product-by-process claims to synthetic cornea or corneal tissue produced by these methods.'[141] Initially, the hearing officer at the UKIPO deemed these patent applications as 'uses of human embryos for industrial or commercial purposes,' falling under paragraph 3(d) of Schedule A2 to the Patents Act 1977, which implements Article 6(2)(c) of the Biotechnology Directive. Consequently, they were considered excluded from patentability.[142] However, upon thorough consideration of the arguments presented by both parties, the CJEU made a significant ruling. According to the CJEU's interpretation of Article 6(2)(c) of Directive 98/44/EC, an unfertilised human ovum, stimulated by parthenogenesis, would not qualify as a 'human embryo' within the context of the provision. This interpretation hinged on whether, based on current scientific knowledge, the ovum possessed the inherent capacity to develop into a human being, a determination to be made by the national court.[143]

As previously discussed, Article 6(2)(c) of the Biotechnology Directive and its reflection in Rule 28(c) of the EPC are essential aspects to consider when examining patent applications related to human pluripotent stem cells, including human embryonic stem cells, and their uses or derived products. A significant milestone in this regard was the public disclosure of a patent application under the PCT (WO 2003/046141) on 5 June 2003, which detailed a procedure for obtaining human parthenogenetic embryonic stem cells from parthenotes (activated oocytes).[144] In light of this, the EPO presently considers that an objection under Article 53(a) and Rule 28(1)(c) EPC as of 1 July 2017 (formerly Rule 28(c) EPC) cannot be raised in respect of

[140] *International Stem Cell Corporation v Comptroller General of Patents, Designs and Trade Marks* [2014] C-364/13.

[141] *Parthenogenic activation of human oocytes for the production of human embryonic stem cells*, Priv L No UK, ch Lifeline Cell Technology, GB0621068D0 (2007) ('*Parthenogenic activation of human oocytes for the production of human embryonic stem cells*'); *Synthetic cornea from retinal stem cells*, Priv L No UK, ch Lifeline Cell Technology, GB0621069D0 (2006) ('*Synthetic cornea from retinal stem cells*').

[142] *Patents Act 1977*s 3(d); Directive (n 29) Article 6(2)(c).

[143] *International Stem Cell Corporation v Comptroller General of Patents, Designs and Trade Marks* [2014] C-364/13 at 39.

[144] *Methods for making and using reprogrammed human somatic cell nuclei and autologous and isogenic human stem cells*, ch Advanced Cell Technology Inc., WO2003046141A2 (2003) ('*Methods for making and using reprogrammed human somatic cell nuclei and autologous and isogenic human stem cells*').

a patent application with subject matter related to human pluripotent stem cells, including human embryonic stem cells, uses thereof or products derived therefrom if:

(i) the application has an effective date (i.e., a valid priority date or, if no priority is claimed or the priority is not valid, a filing date) on or after 5 June 2003, and
(ii) its technical teaching can be put into practice using human embryonic stem cells derived from parthenogenetically activated human oocytes.[145]

Taking into account the EPO's updated patent examination guidelines in March 2022, further clarity has been provided regarding the patentability of subject matter related to human embryonic stem cells and induced pluripotent stem cells.[146] The guidelines stipulate that foetal and post-natal human cells are generally not excluded from patentability. Additionally, culture media, supports, and apparatuses designed for use with human embryonic cells are not automatically excluded either, as their production typically does not require the use of human embryos as base material. Furthermore, the exclusion of using human embryos for industrial or commercial purposes does not hinder inventions meant for therapeutic or diagnostic purposes applied to the human embryo and found beneficial.[147] These developments signify the dynamic nature of patent law as it adapts to advances in biotechnology, particularly in the realm of stem cell research. By providing further clarification and exceptions, the EPO aims to find a fair balance between ethical considerations and promoting innovation and progress in the field of biotechnology.

A patent application involving a bioink utilising human embryonic stem cells may encounter exclusion under Article 53(a) and Rule 28(c) of the EPC if these stem cells are derived from destroyed human embryos. However, if the specific invention could be performed without a destructive use of human embryos, using induced pluripotent stem cells instead, such an invention would be unlikely to be excluded from being patentable.

Patentability of inventions with claims directed towards or including human embryonic stem cells in Australia and the US was discussed in Chaps. 3 and 4 of this book. In Australia, the patentability of human embryonic stem cell-related inventions is governed by s 18(2) of the *Patents Act* which restricts the patentability of human beings and the biological process for their generation. Section 18(2) of the *Patents Act* s more focused on considering a human embryo as a prospective human being and the biological processes involved in its cultivation.[148] Although this perspective is similar to the EPO's decisions on *ordre public* or morality, the

---

[145] *Axiogenesis AG v GE Healthcare UK Limited* [2019] T0385/14; *European Patent Office*, Guidelines for Examination, List of exceptions, G-II 5.3.

[146] "List Of Sections Amended In 2022 Revision—Guidelines for Examination", *Epo.Org* (Webpage, 2022) https://www.epo.org/law-practice/legal-texts/html/guidelines/e/m.htm.

[147] *European Patent Office*, Guidelines for Examination, List of exceptions, G-II 5.3; Directive (n 29) Rec. 42.

[148] See Sect. 3.2 of Chap. 3.

EPO's requirements are more directed towards destruction of embryos which reflects direct morality concerns in comparison with the Australian provisions.

The same perspective is reflected in the analysis provided in 3.2.1 of Chap. 3, where the implications of s 18(2) of the *Patents Act* on inventions related to bioprinted tissues and bioinks would only be applicable if the claimed subject matter is directed towards bioprinting of a human being or a process for their generation.[149] The use of stem cells from an embryo that is unable to progress through additional stages of birth and cannot, therefore, develop into a human being is the only commonality between the Australian and EPO approaches. An invention with claims directed towards a bioprinted tissue or bioink that include embryonic stem cells that are derived from an embryo that cannot give rise to a human being would be considered patentable in Australia and the EPO. On the other hand, the patent system in the US does not provide any specific provisions or restrictions on patentability based on *ordre public* or morality. However, subject matter eligibility standards established by US courts or statutes are applied to determine the patentability of claims directed towards bioprinted tissues and bioinks that include embryonic stem cells.

The patent prosecution analysis provided in Chap. 6 of this book provides an analysis of EU patent applications with claims directed towards bioprinted tissues and bioinks. Chapter 6 analyses how specific EU patent applications have received objections on the grounds of Article 53(a) and Rule 28(c) EPC, and how recent patent applications with subject matter related to bioiprinted tissues and bioinks with subject matter related to pluripotent stem cells and human embryonic stem cells have faced patentability objections based on the provisions of the EPC.

5.4.3 *Methods of Treatment by Surgery or Therapy*

In 1904, a significant ruling by the German Patent Office stated that the 'treatment of the human body (unlike the animal body) can under no circumstances be the topic of an industrially applicable and hence patentable invention.'[150] This decision was grounded in the notion that treatment, being a non-industrial process, does not meet the patentability criteria of an industrial application, even if it involves the discovery of a new therapeutic application for a known substance. The argument against patenting treatment procedures is based on the principle that clinicians must be free to choose their therapy in the interests of public health. Physicians' efforts are directed at maintaining or restoring human health, as well as the protection of life. This raises a crucial question about the suitability of patent law, which primarily deals with industrial property rights, for regulating medical activities and imposing

---

[149] See Sect. 3.2.1 of Chap. 3.

[150] Appeal Department II, Patent Office, *Badewasser*-1905 Bl.f.PMZ 4.

restrictions on them.[151] This stems from a concern that according property rights to methods of medical treatment will result in the living human body becoming a technological object, and patents, licences, and ensuing marketing activities would drive up medical costs that would be contrary to public health interests.[152]

Article 53(c) of the EPC plays a crucial role in defining the scope of patentability concerning medical treatments and diagnostic methods practised on the human or animal body. This provision explicitly excludes such methods from patent protection.[153] However, the same Article further recites that, 'this provision shall not apply to products, in particular substances or compositions, for use in any of these methods.'[154] The interpretation and implications of Article 53(c) have been subjects of extensive analysis and deliberation by the EPO's TBA and EBA.

Medi-physics, Inc. is an important case that illustrates how the provisions related to Article 53(c) of the EPC are analysed and interpreted for patent applications related to methods of medical treatment.[155] In the *Medi-physics* case, the EBA looked at the EPC's legislative history to see whether it gave any indication that only surgical treatments of a curative character were excluded from patent protection under Article 53(c).[156] Before attempting to define the term 'surgery,' the EBA attempted to learn more about the legislative history as well as the intention and purpose of the term 'surgery' being inserted into the EPC.[157] The appeal before the EBA, in this case, concerned the appellant's appeal against the Examining Division's refusal of its European patent application with claims directed towards methods involving magnetic resonance for imaging the pulmonary and/or cardiac vasculature, evaluating blood flow using dissolved polarised 129Xe, and determining a spectroscopic signal representative of a subject's blood volume or flow, including a cardiac imaging method.[158]

The Examining Division had determined that the claimed methods fell under diagnostic methods practised on the human or animal body, rendering them excluded from patent protection under Article 52(4) EPC 1973 [now Article 53(c)].[159] Furthermore,

---

[151] Doris Thums, 'Patent Protection for Medical Treatment—A Distinction Between Patent and Medical Law' (1996) 27(4) *IIC: International Review of Industrial Property and Copyright Law* 423–447.

[152] Eddy D. Ventose, 'Disclaiming methods of medical treatment under the European Patent Convention' (2009) 4(10) *Journal of Intellectual Property Law & Practice* 706–713.

[153] *European Patent Convention*, opened for signature 5 October 1973, (entered in force 7 October 1977), Article 53 (c).

[154] Ibid.

[155] *Medi-Physics, Inc.* G0001/07 [2010].

[156] Ibid., 1.

[157] Ibid., 2.

[158] *MR Methods for imaging pulmonary and cardiac vasculature and evaluating blood flow using dissolved polarized 129Xe*, Priv L No EU, ch Medi Physics Inc, EP1066537A1 (2002) ('*MR Methods for imaging pulmonary and cardiac vasculature and evaluating blood flow using dissolved polarized 129Xe*').

[159] A revised version of the European Patent Convention entered into force on 13 December 2007. The provisions of the revised Convention apply unless the transitional provisions provide for the

the claimed procedures included the process of administering a subject polarised (129) Xe as an imaging agent, either by inhalation or injection. The claimed procedures were prohibited from patent protection under Article 52(4) EPC 1973 [Article 53(c)] as comprising a surgical step since the imaging agent was delivered via injection. The TBA referred the case to the EBA, asking:

> ...Is a claimed imaging method for a diagnostic purpose that includes or encompasses a step involving a physical intervention performed on a human or animal specifically excluded from patent protection as a 'method for treating the human or animal body by surgery' under Article 52(4) EPC, if the step does not per se aim to maintain life and health? If the answer to this question is yes, may the exclusion from patent protection be avoided by amending the claim wording to exclude or disclaim the step-in question, or by allowing the claim to encompass it without being confined to it? Is a claimed imaging method for diagnostic purposes to be considered a constitutive step of a 'treatment of the human or animal body by surgery' if the data obtained by the method allows a surgeon to decide on the course of action to be taken during a surgical intervention immediately...?[160]

In response to the complex questions posed, the EBA conducted a meticulous analysis and arrived at several significant conclusions. Firstly, the EBA determined that the claimed imaging method indeed involved a substantial physical intervention on the human or animal body, requiring specialised medical expertise and entailing considerable health risks. As a result, the EBA categorically classified this method as a 'method for treating the human or animal body by surgery' under Article 53(c) EPC.[161] The EBA provided three grounds for its reasoning and held that a claim that comprises a step that includes an embodiment that is a 'method for treating the human or animal body by surgery' as defined by Article 53(c) EPC cannot be left to encompass that embodiment.[162]

In relation to the second question, the EBA elucidated that the exclusion from patentability under Article 53(c) EPC could be circumvented by explicitly disclaiming the embodiment in question. However, the EBA emphasised that any claim, inclusive of the disclaimer, must fully comply with all the requirements of the EPC, as well as any applicable prerequisites for a disclaimer to be permissible, as established in prior decisions such as *PPG Industries Ohio, Inc. and Genetic Systems Corporation v Roche Diagnostics GmbH & Dade Behring Marburg GmbH*.[163] Moreover, whether altering the claim's wording to exclude the surgical step would be

applicability of the EPC 1973. The provisions provided in Article 52(4) EPC of 1973 are reiterated in Article 53(c) of 2007.

[160] *Medi-Physics, Inc.* T0992/03 [2006].

[161] *Medi-Physics, Inc.* G0001/07 [2010], 1.

[162] Ibid.

[163] *PPG Industries Ohio, Inc.* [2004] G0001/03; *Genetic Systems Corporation v Roche Diagnostics GmbH & Dade Behring Marburg GmbH* [2004] G0002/03. These two similar judgements concern the admissibility of disclaimers in the prosecution of a European patent application or in opposition procedures pertaining to a granted European patent. They address whether and under what conditions an amendment imposing a negative restriction to a claim is allowed where neither the negative limitation nor the subject matter excluded from the claim's scope by it has a basis in the application as filed [Article 123(2) of the EPC].

permissible under the EPC had to be determined on a case-by-case basis, considering the specific circumstances of each claim.[164]

Lastly, the EBA offered clarity on the third question, affirming that the data generated through the imaging technique instantaneously facilitated a surgeon's decision-making during surgical interventions. In light of this, the claimed imaging method was not considered a 'treatment of the human or animal body by surgery' within the context of Article 53(c) of the EPC.[165]

Reflecting the level of advancement of bioprinting technologies, it is evident that methods of using bioprinted products and bioinks may fall within the scope of Article 53(c) of EPC as many applications of bioprinted tissues and bioinks are directly related to diagnosis and treatment of the human or animal body by surgery.[166] One notable example is handheld bioprinting, which has recently gained prominence for its ability to deliver biomaterials and bioinks to injury sites using user-instructed deposition procedures, with a focus on in situ wound dressing and repair.[167] A patent application with subject matter related to this method would certainly fall within the scope of Article 53(c) and be considered a method of treatment. Similarly, a patent application with claims directed towards a method of delivering a bioink through a bioink delivery instrument such as a BioPen[168] directly onto the site of an injury would also fall within the scope of Article 53(c) of EPC.[169]

The patentability of methods of medical treatment in Australia and the US, discussed in Chaps. 3 and 4, provides contrasting perspectives. In Australia, methods of medical treatment are examined under the lens of 'economic significance' and 'industrial applicability.' A claim directed towards a method of handheld in vivo bioprinting used to bioprint a tissue or bioink treating a patient would be considered a medical activity or procedure of doctors (and other medical staff), 'essentially non-economic' and further not 'susceptible' or 'capable' of industrial application, and hence not patentable. On the other hand, the US perspective differs as that jurisdiction does not have any restriction on such subject matter. However, 35 USC 287 (c)(1) of the AIA refuses a remedy for the infringement of such a patented process, effectively nullifying the right in the absence of enforcement. As such, a patent application with claims directed towards a method of handheld bioprinting used to bioprint a tissue or bioink in vivo as a part of treating a patient (including surgical treatment) would be considered patentable in the US. However, even if such a patent application has satisfied the other patentability criteria such as novelty and obviousness and

---

164 *Medi-Physics, Inc.* G0001/07 [2010].

165 Ibid.

166 Thomas H. Jovic et al., '3D Bioprinting and the Future of Surgery' (2020) *Frontiers in Surgery* 129.

167 Hongbin Li et al., 'Handheld bioprinting strategies for in situ wound dressing' (2021) 65(3) *Essays in biochemistry* 533–543.

168 Yu Long Han et al., 'BioPen: direct writing of functional materials at the point of care' (2014) 4(1) *Scientific reports* 1–5.

169 Zahra Pazhouhnia et al., 'Portable hand-held bioprinters promote in situ tissue regeneration' (2022) *Bioengineering & Translational Medicine* e10307.

received a grant, the granted patent may not be enforceable due to the provisions of 35 USC 287(c)(1) of AIA.

5.5 Recommendations and Reflections: Patenting Framework Model—Europe

Patentable subject matter requirements relevant to bioprinted tissues and bioinks in the EU are primarily governed by Articles 2, 5, 6, and 8 of the EU Biotechnology Directive and Articles 52, 53(a), and 53(c) of the EPC.[170] Many aspects of the Biotechnology Directive are merely declaratory of pre-existing national and European Patent Office (EPO) practices. For example, Article 6(1) of the Directive and Recitals 36–39, 43, and 44 mirror the spirit of Article 53(a) of the EPC; whereas Article 6(2) (in tandem with Recitals 40–42) gives a non-exhaustive list of unpatentable subject matter based on ethical and moral grounds.[171] The analysis contained in Sect. 5.4 of this chapter sheds light on the application of different Articles of the Biotechnology Directive that may be relevant to bioprinted tissues and bioinks. Section 4 (according to the provisions of Articles 2 and 5 of Biotechnology Directive) identified that the structure and composition of bioprinted tissues may be similar to that of natural tissues and their composition; however, as bioprinted tissues and bioinks are produced by a technical process, they may be considered to be within the scope of patentable subject matter.[172]

The analysis of Article 8 of the Biotechnology Directive indicated that a patent application with claims directed towards a bioink may have an extended scope of protection if the claimed bioink is used to develop an identical bioink (possessing the same characteristics) through propagation or multiplication.[173] Similarly (considering the provisions of Article 8(2)), in relation to a patent application with claims directed towards a process of bioprinting a tissue or a process of formulating a bioink, such protection may extend to the biological material directly obtained from that process and any subsequent generation of biological material derived from such material which possesses the same characteristics as the first-generation biological material.[174] If a bioprinted tissue is developed using a particular bioprinting process, patent protection may also extend to the bioprinted tissue developed using the patented bioprinting process. This indicates that if a patent applicant only claimed a process of developing a bioprinted tissue and not the bioprinted tissue itself, according to the provisions of Article 8(2) of the Biotechnology Directive, patent protection

[170] See Sects. 5.3 and 5.4 of this chapter.

[171] See Sect. 5.3.1 of this chapter.

[172] Ibid.

[173] Ibid.

[174] Ibid.

may extend to the unclaimed bioprinted tissue only because it can only be derived from a patented bioprinting process.[175]

Recommendations (Europe)

The following strategies are recommended to patent applicants to satisfy the patentable subject matter requirements for inventions with claims directed towards bioprinted tissues or bioinks.

i. A claim directed towards a bioprinted tissue or bioink that includes any elements isolated from the human body (such as stem cells or biomolecules or enzymes), or otherwise produced by a technical (bioprinting) process, may satisfy the requirements of Article 52(2) of the EPC, and be considered patentable.
ii. In light of the *order public* or morality exception (Article 53(a) of EPC), it is recommended that inventors who seek to patent their inventions in Europe develop bioprinted tissues and bioinks using stem cells (for example, induced pluripotent stem cells) other than embryonic stem cells.
iii. Considering Article 53(c) of EPC, a patent claim directed towards a method of bioprinting a tissue or a bioink directly on the injured site of a human or animal body may be objected as it would be considered as a method of medical treatment.

Article 52(2) of the EPC precludes from patentability discoveries, scientific theories, mathematical methods, aesthetic creations, schemes, rules, and methods for performing mental acts, playing games or doing business, computer programs, and presentations of information from patenting 'as such.'[176] As one of the major essential elements of bioprinted tissues and bioinks is stem cells, subject matter directed towards bioprinted tissues and bioinks may be excluded as discoveries in light of Article 52(2) if the claims are directed towards stem cells 'as such.'[177] However, exclusion of this subject matter may not be in question if a claim directed towards bioprinted tissues and bioinks includes stem cells, and the claim as a whole has a practical effect (such as the effect of specific bioprinted tissue or bioink), resulting in a 'technical effect.'[178] Section 5.1 further provided analysis of case law relevant to Article 52 of the EPC and concluded that claims directed towards a bioprinted tissue may satisfy the requirements of Article 52 of the EPC.[179]

Further, in contrast to the US and Australian patent provisions, the patent provisions of the EPC have a major focus on *ordre public* or morality in determining the patentability of bioprinting subject matter due to the use of stem cells in bioprinted

[175] Ibid.

[176] See Sect. 5.4.1 of this chapter.

[177] Ibid.

[178] Ibid.

[179] Ibid.

tissues and bioinks.[180] Article 53(a), in light of Rule 28(c), excludes specific subject matter from patentability if it includes the use of human embryos for industrial or commercial purposes.[181] Under the same provisions, subject matter claiming human embryonic stem cells derived by a process resulting in the destruction of the human embryos are excluded from patentability.[182] However, if a specific invention could be performed without presupposing a destructive use of human embryos which would not contravene Article 53(a) and Rule 28(c) of the EPC, the subject matter may not be excluded.[183] If a patent application with claims directed towards bioprinted tissues or bioinks that uses a different type of stem cells (such as induced pluripotent stem cells), such patent claims may also not fall within the scope of Article 53(a) and Rule 28(c) of the EPC.[184] The analysis here also focused on patent claims directed towards methods of treatment by surgery or therapy. Patent applications with claims directed towards methods of bioprinting tissues and bioinks may fall within the scope of exclusions in Article 53(c) of the EPC, if specific subject matter includes methods of treatment of the human or animal body by surgery.[185] For example, a patent application with subject matter related to a method of bioprinting a tissue or a bioink directly on the injured site of a human or animal body would fall within its scope as such a method would be considered a method of medical treatment.[186]

The analysis in this chapter addressed whether the patentable subject matter requirement is a hurdle to patenting bioprinting inventions in the form of bioinks and bioprinted tissues, and the applicability of the subject matter requirements and doctrinal limits to 3D bioprinted tissues and bioinks from the European perspective. In light of the insights of the European analysis, the answers to the research questions are identified to be two-fold. Firstly, considering recent updates to the examination guidelines of the EPO, subject matter related to bioprinted tissues or bioinks that includes embryonic stem cells would not be excluded from patentability in light of Article 53(a) and Rule 28(c) of EPC.[187] Secondly, patent applications with claims directed towards methods of bioprinting a tissue or bioink on the human or body by surgery may not be considered patentable as it would fall within the scope of Article 53(c) of the EPC.[188] Although, on the whole, the patentable subject matter requirement is not a hurdle to patenting bioprinting inventions, the opposite conclusion may be reached where Article 53(c) EPC is applicable.[189] Again, the analysis in Chap. 6 was intended to tease out the factual discrepancies that determine patentability under EU law. Figure 5.1 provides a patentability framework model and

[180] See Sect. 5.4.2 of this chapter.

[181] Ibid.

[182] Ibid.

[183] Ibid.

[184] Ibid.

[185] See Sect. 5.4.3 of this chapter.

[186] Ibid.

[187] See Sect. 5.4.2.1 of this chapter.

[188] See Sect. 5.4.3 of this chapter.

[189] Ibid.

identifies the types of patent claims that are patentable with respect to the subject matter in question. Adherence to the recommendations is intended to overcome the challenges to patenting relevant bioprinting inventions outlined in this book.

5.6 Conclusion

The European patent regime is governed by multiple legal and regulatory layers in which the EPC plays a major role. In addition, the Biotechnology Directive provides a vital direction to the patenting of inventions with subject matter directed towards biological materials and technologies. To obtain a patent for bioprinted tissues and bioinks, the application must fulfil various patentability criteria set by the EPC, which differ from those in other jurisdictions. Notably, subject matter claiming product and process for bioprinted tissues and bioinks must navigate the patentability exclusions in Articles 53(a) and 53(c) of the EPC.

Most of the patent applications with subject matter related to bioprinted tissues and bioinks include different types of stem cells, because these are core elements of bioprinted tissue or bioinks. However, Article 53(a) in light of Rule 28(c) excludes specific subject matter from patentability if it includes the use of human embryos for industrial or commercial purposes. As provided in the case analysis in Sect. 5.4 of this chapter, the TBA and EBA have analysed this in different instances and concluded that subject matter claiming human embryonic stem cells derived by a process resulting in the destruction of the human embryos are excluded from patentability. However, if a specific invention could be performed without presupposing a destructive use of human embryos, the subject matter may not be excluded from being patentable. The EPO examination guidelines reflect these provisions by outlining how subject matter related to embryonic stem cells may be considered patentable if the claimed stem cells are acquired without presupposing a destructive use of human embryos. For example, a procedure for obtaining human parthenogenetic embryonic stem cells from parthenotes (activated oocytes) is considered by the EPO to be an acceptable method of acquiring embryonic stem cells from human embryos. Thus, if a patent application has claims directed towards a bioprinted tissue or bioink that is comprising parthenogenetic embryonic stem cells, such claims may be considered patentable under these recent examination guidelines.

Patent applications with claims directed towards methods of bioprinted tissues and bioinks may fall within the scope of exclusions provided in Article 53(c) of the EPC, if specific subject matter includes methods of treatment of the human or animal body by surgery. A method of bioprinting directly onto a wounded site would fall within this exclusion, being considered a surgical treatment. The evolving European patent system has experienced profound changes over recent decades, and further adaptations may be necessary to accommodate advancements like bioprinted tissues and bioinks.

Considering the recent outcomes of directives, regulations, and judicial analysis, it is evident that the European patent system has undergone a profound transformation

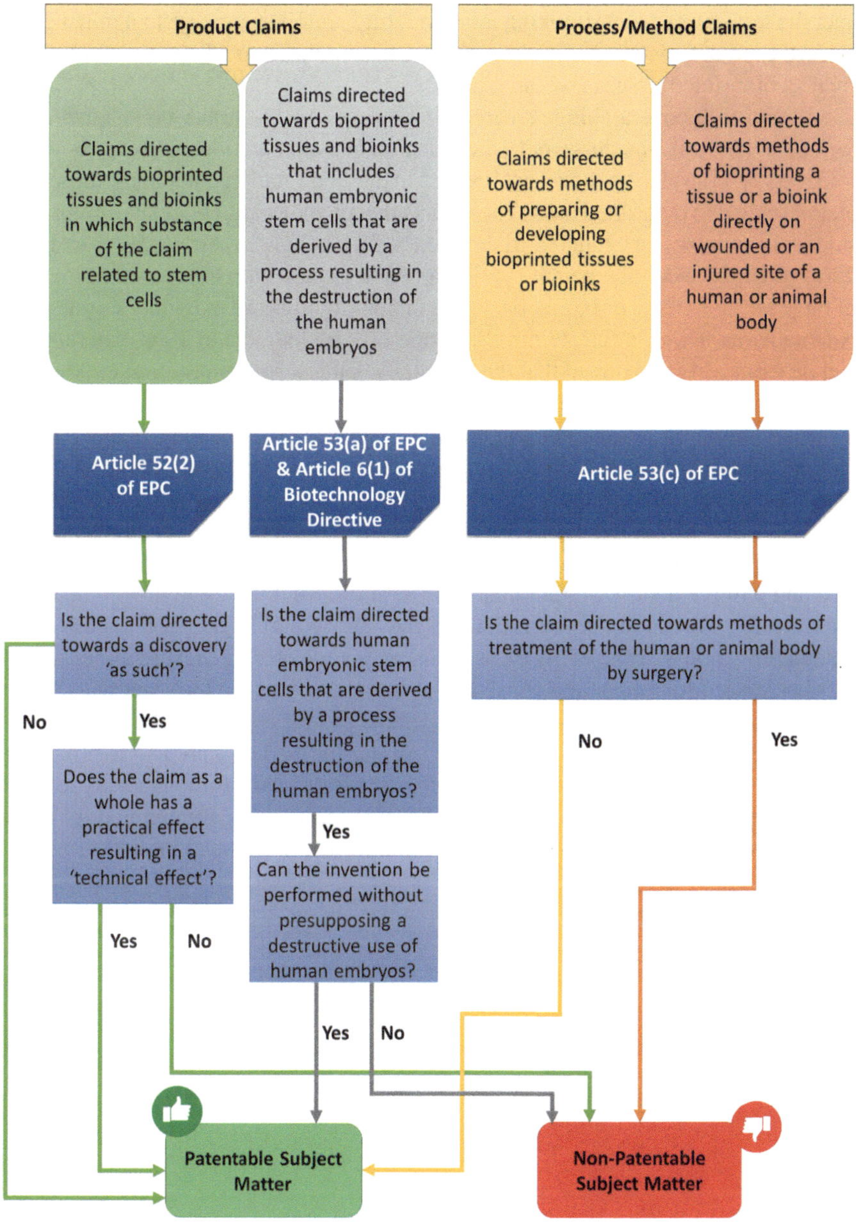

Fig. 5.1 Visualisation of proposed patentability framework model for patenting inventions related to bioprinted tissues and bioinks in the Europe

over the last few decades. Looking into the future, the EU patent system may need to be further transformed to adjust to subject matter related to advanced technologies such as bioprinted tissues and bioinks.

In light of the overarching research question of this book and the insights of the European analysis, the answers to the research question are identified to be in two-fold. Firstly, considering the recent updates in the examination guidelines, the subject matter related to bioprinted tissues or bioinks that include parthenogenetic embryonic stem cells may not be excluded from patentability in light of Article 53(a) and Rule 28(c) of EPC. Secondly, patent applications with claims directed towards methods of bioprinting a tissue or bioink on a human body by surgery may not be considered patentable as it would fall within the scope of Article 53(c) of EPC. Therefore, it can be tentatively concluded that the patentable subject matter requirement does not appear to be a fundamental hurdle to patenting bioprinting inventions. At this point, it is also important to note that the answers to these questions may vary on a case-to-case basis and be based on how specific patent applications claiming bioprinted tissues or bioinks are drafted.

Bioprinting technologies are emerging as a promising field of innovation for medical treatments. However, they also pose legal challenges for patent protection and enforcement. Different countries have different views on how to deal with these issues. For example, Australia and the US have divergent approaches to the patentability and enforceability of bioprinting methods for medical treatments. This shows the need for a clear and consistent patent framework for bioprinting technologies. A fair framework that supports innovation, protects public health, and respects patent rights is important to promote bioprinting for medical care and treatment. In light of this, the next chapter (Chap. 6) will provide an analysis of patent prosecution data to identify how the above-discussed provisions are actually applied by patent examiners on patent applications claiming bioprinted tissues and bioinks.

Chapter 6
Evaluation of Patent Prosecution Data of Patent Applications Claiming 3D Bioprinted Tissues and Bioinks

6.1 Introduction

Patent prosecution data is crucial in understanding how the current patent provisions and case law are interpreted and applied. Considering the importance of patent prosecution data and the scope of this book, the present chapter reports the results of an empirical analysis using the publicly available patent prosecution data from patent applications with claims directed to bioprinted tissues and bioinks in Australia, the US, and the EU. The insights provided in this chapter will also contribute towards answering the overarching research question of this book, and will also address some of the subsidiary research questions, particularly 'What types of patent claims are being applied for and granted/objected to in relation to 3D bioprinted tissues and bioinks?' and 'How are patent offices currently assessing and responding to patent claims related to 3D bioprinted tissues and bioinks?' This chapter provides insights into how patent examiners are interpreting and applying the patent provisions related to patentable subject matter and other patentability criteria to accept or object to specific subject matter related to bioprinting tissue and bioinks. A comparative approach has been employed in analysing the results and insights of the patent prosecution data analysis. Considering the research questions and the scope of this book, the patent prosecution data analysis is primarily focused on patentable subject matter requirements, with a briefer overview of the other patentability criteria, these being novelty, inventive step, utility, and the disclosure requirements.

Patent applications around the world must go through a stringent examination procedure before they can be granted. During the examination procedure, patent examiners assess a patent application in light of patentability criteria: patentable subject matter, novelty, inventive step, utility, and disclosure and will only grant a patent if the subject matter disclosed in a patent application satisfies the patentability

P. Devarapalli, *Bioinked Boundaries*, https://doi.org/10.1007/978-981-96-2591-8_6

criteria.[1] The proceedings between the patent office and the patent applicant from filing of the application until the grant of a patent is part of the so-called 'patent prosecution history.'[2] Most patent offices around the world publish all the documents that are exchanged between the patent office and patent applicant during the process of patent examination. The documents that are part of the patent prosecution history include published patent applications, examination reports or office actions (providing the details of objections raised by the patent examiner in light of their examination of the patentability criteria), applicant responses to examination reports, and amendments made by the applicant in response to objections. Patent applications contain a number of claims that encompass the invention intended to be protected. However, during the patent examination procedure, the patent examiner may object to some of the patent claims in light of specific patentability criteria. The examination report or office action also provides details of these grounds of objection, affected claims, and the related prior art.

The data provided in the patent prosecution history is crucial to understand how patent examiners interpret and apply the current statutory provisions and case law to raise objections on patentability of diverse types of subject matter disclosed in the patent applications. Specifically, the published data that is part of patent prosecution is especially valuable for empirical research focusing on the patent examination process, examiner heterogeneity, and application and litigation outcomes.[3] Different empirical studies have been based on the publicly available patent prosecution data and these provide insights on the basis of grounds of patent rejections, patent examiner characteristics, and office action traits.[4]

---

[1] Giles S. Rich, 'Principles of Patentability' (1959) 28 *George Washington Law Review* 393; John F. Duffy, 'Rules and Standards on the Forefront of Patentability' (2009) 51 *William & Mary Law Review* 609.

[2] David Vaver, 'The Perplexities of Patent Prosecution History: Procedure over Principle?' (2021) *Osgoode Legal Studies Research Paper Forthcoming.*

[3] Iain M. Cockburn, Samuel S. Kortum and Scott Stern, 'Are all patent examiners equal? The impact of examiner characteristics' (Pt National Bureau of Economic Research Cambridge, Mass., USA) (2002). National Research Council, *Patents in the knowledge-based economy* (National Academies Press, 2003); Douglas Lichtman, 'Rethinking prosecution history estoppel' (2004) *The University of Chicago Law Review* 151–182. Ronald J. Mann, 'The Idiosyncrasy of Patent Examiners: Effects of Experience and Attrition' (2013) 92 *Tex. L. Rev.* 2149. Michael Carley, Deepak Hedge and Alan Marco, 'What is the probability of receiving a US patent' (2015) 17 *Yale JL & Tech.* 203. Michael D. Frakes and Melissa F. Wasserman, 'Does the US patent and trademark office grant too many bad patents: evidence from a quasi-experiment' (2015) 67 *Stan. L. Rev.* 613. Ronald J. Mann, 'The Idiosyncrasy of Patent Examiners: Effects of Experience and Attrition' (2013) 92 *Texas Law Review* 2149.

[4] Mark A. Lemley and Bhaven Sampat, 'Examiner characteristics and patent office outcomes' (2012) 94(3) *Review of economics and statistics* 817–827. Qiang Lu, Amanda Myers and Scott Beliveau, 'USPTO patent prosecution research data: Unlocking office action traits' (2017) *Economic Working Paper Series.*

6.2 Materials and Methods

An empirical research methodology was designed to analyse the patent prosecution data of patent applications claiming bioprinted tissues and bioinks published in the AusPat database provided by IP Australia, the Public Pair and Office Action Research Dataset for Patents (OARDP) databases provided by United States Patent and Trademark Office (USPTO), and the European Patent Register provided by the European Patent Office (EPO). The methodology is elaborated below in detail for each of the different steps.

6.2.1 Patent Search Strategy and Screening

A patent search strategy was developed to retrieve patent applications with claims directed towards bioinks and bioprinted tissues. The search strategy was developed using a number of keywords and their synonyms (bioink, biogels, hydrogel, bioprint, tissue, scaffold, structure, organ, cell) to ensure all the relevant applications were retrieved.[5] The search was specifically narrowed to claims in order to identify patent applications that specifically claim bioinks and bioprinted tissues. This search strategy was used because a patent search directed towards the complete patent description would retrieve all the patent applications having 'bioinks' and 'bioprinted tissues' and similar words which would include false positive patent applications that mention 'bioinks' and 'bioprinted tissues' in the patent description or background but do not have claims directed towards 'bioinks' and 'bioprinted tissues.'

The search was conducted in multiple open access patent databases such as Google patent, Patent Lens, Patentscope, and Espacenet and subscription-based patent databases such as Derwent innovation and Patseer to ensure all relevant patent data was covered.[6] These databases collate and organise the publicly available patent data as provided in the patent registers of different jurisdictions. The access to subscription-based patent databases was acquired on demonstration and trial basis.

These databases have their own collection of patent applications, and patent applications that are available in one database may not necessarily be available in another database. Hence, patent searches were conducted using the designed search strategy

---

[5] Search strategy: (((bioink) OR (bio-ink) OR (biogel) OR (bio-gel)) OR ((bioprint* OR bio-print*) AND tissue)) OR ((bioprint* OR bio-print*) AND scaffold)) OR ((bioprint* OR bio-print*) AND structure)) OR ((bioprint* OR bio-print*) AND organ)) OR ((bioprint* OR bio-print*) AND cell)).

[6] Alireza Noruzi and Mohammadhiwa Abdekhoda, 'Google Patents: The global patent search engine' (2014) 11(1) *Webology*; Björn Jürgens and Victor Herrero-Solana, 'Espacenet, Patentscope and Depatisnet: A comparison approach' (2015) 42 *World Patent Information* 4–12; Lens.org. 2021. *The Lens—Free & Open Patent and Scholarly Search*. [online] Available at: https://www.lens.org/lens/search/patent [Accessed 24 November 2021]; Derwent. 2021. *Derwent Innovation—Derwent*. [online] Available at: https://clarivate.com/derwent/solutions/derwent-innovation/ [Accessed 24 November 2021]; PatSeer. 2021. *PatSeer*. [online] Available at: https://patseer.com/ [Accessed 24 November 2021].

in all of the above databases to ensure all relevant patent applications were retrieved. The patent applications that were retrieved using the designed search strategy were further refined to identify the patent applications filed in the US, EU, and Australia. This step was performed using refining tools that identify patent applications filed in specific jurisdictions. Once the relevant patent applications were retrieved, they were combined in a spreadsheet and individual patent applications were manually reviewed and screened to exclude irrelevant patent applications and duplications. Following screening, a total of 342 patent applications were finalised for further patent prosecution analysis, of which 80 were published by IP Australia, 164 by the USPTO, and 97 by the EPO.

6.2.2 Retrieval and Analysis of Patent Prosecution Data

The final list of patent applications was further classified according to their jurisdiction. Their patent prosecution history was examined to identify the specific types of claims that were objected to, and the type of objection. This step was performed using the patent prosecution database for particular jurisdictions (Public Pair for US patents, AusPat for Australian patents, and the European Patent Register for EU patents).

The patent prosecution analysis was focused on patent applications that were currently in the prosecution stage and had received at least the first examination report. The first examination reports provide the details of claims that are initially objected to, based on patentability requirements. These reports are thus helpful for this study to identify the types of claims that have been objected to, based on the patentability criteria of specific jurisdictions.

6.2.3 Spreadsheet Analysis

Individual spreadsheets were created for each jurisdiction. In the spreadsheet, each column was designated with specific data associated with an individual patent application. For example, the spreadsheet for Australian patent applications included columns designated with publication number, title, independent claims, legal status, types of claims, patentable subject matter, novelty, inventive step, and disclosure requirements such as fair basis, full description, clarity, and lack of unity. In the next stage, patent publication numbers were used to retrieve the prosecution data of relevant patent application. Once the patent prosecution data of specific patent application was retrieved from the relevant database (AusPat for Australia, Public Pair for the US, and the European Patent Register for EU), it was manually added to the designated column in the spreadsheet. For example, if a first examination report of a patent application disclosed that the examiner raised objections for claims directed towards bioinks on the basis of patentable subject matter requirements, the same data was manually entered in the column designated for patentable subject matter for the specific patent application. Using this approach, the spreadsheet analysis for all the identified patent applications of Australia, the US, and the EU was completed.

6.2.4 Data Analysis and Visualisation

The patent prosecution data provided in the spreadsheets was further analysed to identify the patent prosecution data insights in each jurisdiction. This step was performed using a text mining, refining, and analysis software known as VantagePoint.[7] VantagePoint is a professional-grade desktop text mining program that provides analysts with a comprehensive set of refining, analysing, and reporting options for scientific, technical, market, and patent data.[8] The spreadsheets containing manually analysed patent prosecution data were imported to VantagePoint and the program was used to identify and sort the data entered in a designated column into the form of tables and graphs. The patent prosecution analysis data that was retrieved from VantagePoint was then exported in the form of Microsoft Excel tables and the same files containing the refined and finalised patent prosecution analysis data were processed in Microsoft Excel and PowerPoint programs to develop visualisations and graphs for comparative analysis.

6.3 Results

Patent prosecution analysis was undertaken for a total of 341 patent applications (Australia: 80, US: 164, EU: 97) with claims directed towards bioprinted tissues and bioinks. This analysis revealed multiple insights into how patent claims related to bioprinted tissues and bioinks are examined by patent offices in Australia, the US, and the EU. The results of the patent prosecution analysis are classified, interpreted, and presented here in six different categories: legal status, types of claims, patentable subject matter, novelty, inventive step, and disclosure and support requirements.

6.3.1 Legal Status

Analysis of the legal status of retrieved patent applications was conducted by considering the various stages of the patent life cycle. Specifically, patent applications that

---

[7] Thevantagepoint.com. 2021. *The VantagePoint*. [online] Available at: https://www.thevantagepoint.com/ [Accessed 24 November 2021].

[8] Ibid.

were granted and maintained at the time the analysis was undertaken were classified as 'Active,' patent applications that were pending for examination or currently in the examination stage were classified as 'Pending,' and patent applications that were abandoned, ceased, expired, and withdrawn were classified as 'Abandoned/ Ceased.' Patent applications classified as active were those that had been granted by the Australian, US, or EU patent offices with maintenance or renewal fees paid on time. Patent applications that were published and in the prosecution stage or examination stage were marked as pending. Patent applications that had received a final rejection from the patent offices, had completed the term of 20 years, or that were not renewed in the required timeframe were classified under the abandoned or ceased category.

Further analysis of the legal status of patent applications filed in Australia, the US, and the EU, as provided in Fig. 6.1, reveals a number of broad trends. In particular, it is obvious that more of the patent applications filed in the EU (67%) and the US (66%) are active in comparison with active patent applications in Australia (30%). In contrast, patent applications filed in Australia have the highest pendency rate (48%), in comparison with patent applications filed in the EU (21%) and US (9%). The EU has the lowest abandonment and cease rate of 12% in comparison with Australia (22%) and the US (25%).

Timely renewal or maintenance of patent applications depends on multiple factors. Demand and commercial viability of technology associated with a specific patent application in a specific jurisdiction is one of the factors that plays a crucial role to decide on timely maintenance of a patent application. It can be inferred that patent applications with claims directed towards bioprinted tissues and bioinks are more likely to be active in the EU and the US. A patent analytics report on stem cell technologies underlying regenerative medicine published by IP Australia in December 2019 provided similar insights on the legal status of patent applications on mammalian pluripotent stem cells. That patent analysis report indicated that:

> 5250 (96 per cent) patent families out of 5447 filed since 2008, relating to mammalian pluripotent stem cell technology, are granted or with protection being sought. Of these, 2289 (42%) patent families have been granted, while 2961 (54.3%) families have patent protection being sought. The remaining 197 (3.6%) families are 'dead', having either lapsed or expired without being granted.[9]

As patent applications related to bioprinted tissues and bioinks are broadly classified under stem cell technology and regenerative medicine, the patent application numbers (48%) (related to bioinks and bioprinted tissues) that are pending for examination in Australia are similar to the percentage (54.6%) of pending patent application data provided in the report.

[9] IP Australia, *Patent Analytics on Stem Cell Technologies Underlying Regenerative Medicine*, Available at: https://www.ipaustralia.gov.au/tools-resources/publications-reports/patent-analytics-stem-cell-technologies-underlying-regenerative [Accessed 18 January 2022].

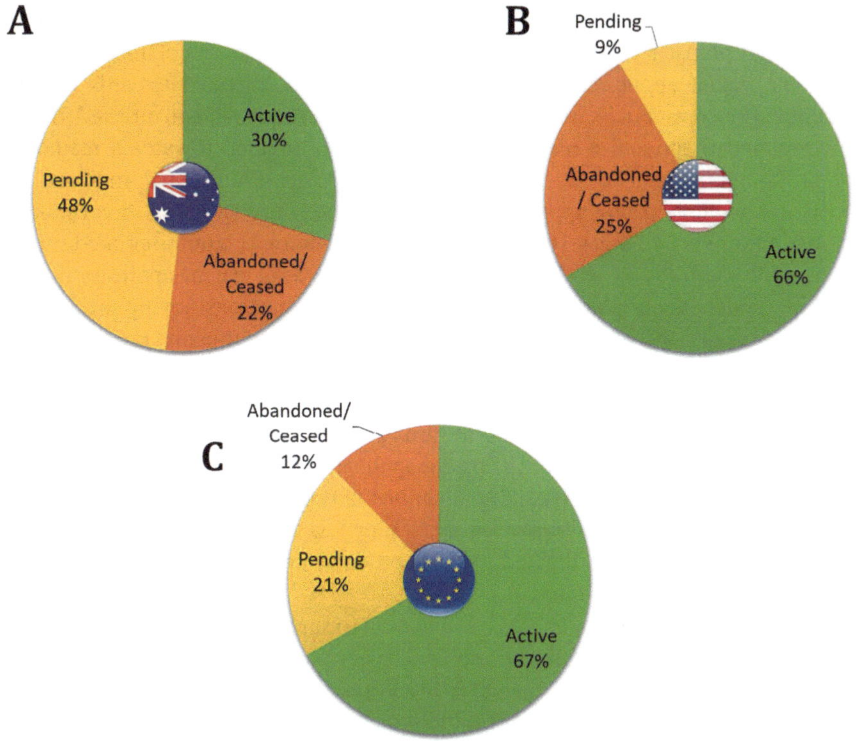

Fig. 6.1 A comparative visualisation of legal status of patent applications claiming bioprinted tissues and bioinks in **A** Australia, **B** United States, and **C** European Union

6.3.2 *Types of Claims*

The claims in a patent or patent application define the scope of the protection provided therein in technical terms. In other words, the purpose of the claim is to specify what subject matter is covered by the patent. Patent claims are classified into two types, independent claims and dependent claims. Independent claims are those that do not refer to any other claim, whereas dependent claims refer to one or more previous claims.[10] Patent claims related to bioprinted tissues and bioinks are further classified as product, process or method, and composition claims. A claim directed towards a bioprinted tissue or organ or scaffold is classified as a product claim whereas a claim directed towards a process or a method of bioprinting a tissue or organ or scaffold is classified as a process or method claim. Similarly, as bioinks are composed of stem cells, biomaterials or biopolymer gels, bioactive molecules, and growth factors,

[10] Alan C. Marco, Joshua D. Sarnoff and AW Charles, 'Patent claims and patent scope' (2019) 48(9) *Research Policy* 103790.

claims directed towards bioinks are mostly classified as composition claims and the methods of preparing bioinks are classified as method or process claims.[11]

Some patent applications have independent claims directed towards multiple inventive concepts. For example, they might include a product claim directed towards a bioprinted tissue and a method or process claim directed towards a method of bioprinting a tissue. The claim type analysis, as provided in Fig. 6.2, indicates that 55% of Australian patent applications claim methods or processes of bioprinting or preparing bioinks whereas 16% of the patent applications claim bioprinted tissues, organs, or scaffolds as products. The remaining 29% of patent applications have multiple independent claims with 15% of patent applications having product and process claims, 10% of patent applications having composition and process claims, and 4% of patent applications having product, process, and composition claims. On the other hand, the US (58%) and the EU (61%) have quite different ratios of patent applications having claims directed towards product and process combinations. In addition, 23% of US and 9% of EU patent applications only claim products whereas 17% of US and 7% of EU patent applications only claim processes or methods of bioprinting or preparing bioinks. Lastly, 1% or less of the US and the EU patent applications have claims directed towards product, process, and composition in a single patent application.

This claim type analysis acts as a foundation to support further analysis of patent prosecution data and individual claim analysis of patent applications that received objections based on different patentability criteria. For example, the above claim type analysis has provided detail on the types of patent claims related to bioinks and bioprinted tissues. The further patent prosecution data analysis and individual claims analysis reveals how these product and process claims related to bioinks and bioprinted tissues were examined and objected to, based on jurisdiction-specific patentability requirements.

6.3.3 Patentable Subject Matter

Subject matter that is patentable, statutory, or patent eligible is that which is acceptable for patent protection. Patent provisions and case law in different jurisdictions provide exclusions from patentability for specific subject matter, even if the invention is novel and non-obvious. A detailed doctrinal analysis of patentable subject matter requirements of Australia, the US, and the EU considering patent claims directed towards bioprinted tissues or organs has been provided in the previous chapters of this book.[12] The present section will focus on providing a comparative analysis of patentable subject matter objections (in Australia, the US, and the EU) resulting from patent prosecution data analysis. In summary, patentable subject matter-based

[11] *Bioink compositions and methods of preparing and using the same*, ch Wake Forest University Health Sciences, (2019) ('*Bioink compositions and methods of preparing and using the same*').

[12] See Chaps. 3, 4, and 5.

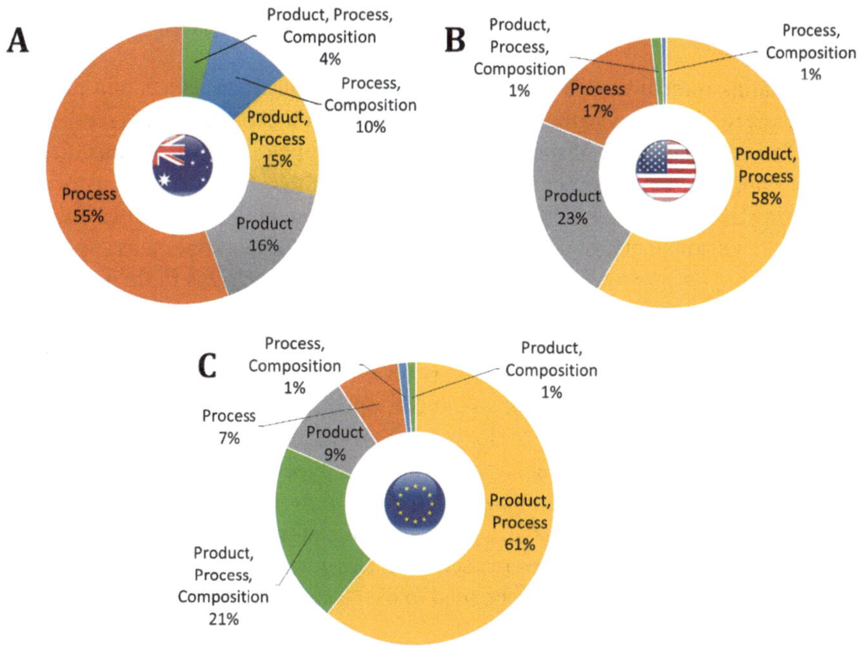

Fig. 6.2 A comparative visualisation of types of claims in patent applications claiming bioprinted tissues and bioinks in **A** Australia, **B** United States, and **C** European Union

analysis of patent prosecution data indicates that some patent applications filed in Australia have received objections from patent examiners for product and process claims directed towards bioprinted tissues or organs, hydrogels, biogels, and cell isolation kits. However, the same analysis on US patent applications reveals very few objections for product and process claims directed towards bioprinted tissues or organs. In contrast, the EU has the highest patentable subject matter objections for product and process claims directed towards bioprinted tissues or organs, bioinks, hydrogels, and bioprinters (Table 6.1).

Table 6.1 A comparative overview of types of patent claims related to bioprinting that received objections based on patentable subject matter requirements of Australia, the US, and the EU

Types of patent claims	Number of patent applications that received patentable subject matter objections		
	Australia	US	EU
Bioprinted tissue/organ and methods	–	2	5
Bioinks, hydrogels, biogels and methods	3	–	3

6.3.3.1 Objections on Bioprinted Tissues/Organs and Methods

Patent claims directed towards bioprinted tissues/organs as products and methods have received objections in the US and the EU, but not in Australia. This section will focus on further analysis of the individual patent applications with claims directed towards bioprinted tissues/organs as products and methods that are objected to by the US and the EU patent examiners on the grounds of patentable subject matter. Table 6.1 illustrates that patent claims directed towards bioprinted tissues or organs as products and processes have received patentable subject matter objections in the US and the EU. In contrast, the patent prosecution data from Australian patent applications does not reveal any objections for claims directed towards bioprinted tissues or organs as products and process. In comparison with the US (two patent applications), the EU has the greatest number of patentable subject matter objections (five patent applications) within this category. Table 6.2 provides a comprehensive overview of patent applications that have received patentable subject matter-based objections in the US, the EU, and Australia. In the list of applications that received subject matter objections as provided in Table 6.2, a group of applications containing similar subject matter from each jurisdiction were identified and further examined to understand how the patent applicants have responded to overcome the patentable subject matter objections.

US

The US patent application entitled 'Three-dimensional bioprinted pancreatic tumor model' having a claim directed towards bioprinted tissue or organ received an objection based on 35 USC 101 (of AIA) patentable subject matter grounds.[13] The objected patent claim is as follows.

> A three-dimensional, engineered, pancreatic tumor model comprising:
>
> a. a stromal microenvironment, the stromal microenvironment comprising pancreatic stellate cells and endothelial cells; and
> b. tumor tissue, the tumor tissue comprising pancreatic cancer cells, the tumor tissue encased in the stromal microenvironment to form the three-dimensional, engineered, pancreatic tumor model;
>
> provided that the stromal microenvironment and the tumor tissue were bioprinted.[14]

The USPTO patent examiner responded that,

> Claim(s) 1-8 is/are directed to products of nature. The claim(s) does/do not include additional elements that are sufficient to amount to significantly more than the judicial exception because the nature based product recited in the claims is not significantly different in structure or function from the structure or function of the naturally occurring product counterparts.

[13] *Three-dimensional bioprinted pancreatic tumor model*, ch Oregon Health Science University, US20160040132A1 (2016) ('*Three-dimensional bioprinted pancreatic tumor model*').

[14] Ibid.

Table 6.2 An overview of patent applications filed in US, EU, and Australia that received objections for claims directed towards bioprinted tissues/organs and bioinks as products and processes

S. no.	Jurisdiction	Application no.	Title of invention	Details of patentable subject matter objection
1	US	US20150050686A1	Omnidirectional, multiaxial bioprinted tissue system, techniques and applications	Judicial exception: claims drawn to a natural product of vascularised tissue
2		US20160040132A1	Three-dimensional bioprinted pancreatic tumor model	Judicial exception: claims drawn to naturally occurring cells
3	EU	EP3463822A1 (also filed in US but no PSM based objection received)	Preparation and applications of RGD conjugated polysaccharide bioinks with or without fibrin for 3D bioprinting of human skin with novel printing head for use as model for testing cosmetics and for transplantation	Article 53(c): Claim 11 relates to methods for treatment of the human or animal body by surgery
4		EP2861270A1 (also filed in US but no PSM-based objection received)	Engineered three-dimensional connective tissue constructs and methods of making the same	Article 53(a) and Rule 28(c): Claim 4 relates to human totipotent cells that are excluded from patentability
5		EP3720386A1 (also filed in US and Australia but pending for examination)	Multi-layer skin constructs and methods of making and using the same	Article 53(c): Claims 11 and 12 relate to methods for treatment of the human or animal body by surgery
6		EP3771547A1	Method for 3D printing of vascularized tissues and organs	Article 53(c): Claim 9 relates to methods for treatment of the human or animal body by surgery
7		EP3471733A1 (also filed in US but no PSM-based objection received)	Reagents, compositions and methods for improving viability and function of cells, tissues, and organs	Article 53(c): Claims 4–10 and 16 relate to methods for treatment of the human or animal body by surgery

(continued)

Table 6.2 (continued)

S. no.	Jurisdiction	Application no.	Title of invention	Details of patentable subject matter objection
8		EP3074415A1 (also filed in US but no PSM-based objection received)	Novel ultrashort hydrophobic peptides that self-assemble into nanofibrous hydrogels and their uses	Article 53(a) and Rule 28(c): Claim 12 relates to hydrogel biofabrics which may comprise stem cells, including human pluripotent (embryonic) stem cells. This objection has been reversed considering decision G2/06. Further analysis added in Sect. 6.3.3.2
9		EP3538169A1	Bioprinted hair follicles and uses thereof	Article 53(c): Claim 10 relates to methods for treatment of the human or animal body by surgery
10		EP3234108A1	Differentiation of pluripotent stem cells to form renal organoids	Article 53(a) and Rule 28(c): Claims relate to hydrogel biofabrics which may comprise stem cells, including human pluripotent (embryonic) stem cells. This objection has been reversed considering decision G2/06. Section 6.3.3.2 of this chapter provides further analysis on this patent application
11	Australia	AU2007327100B2 (also filed in the US and the EU but no PSM-based objection received)	Biogel	Claims 1–15 are directed at a non-patentable collocation of known agents as they comprise known biomolecule immobilised carriers and known compounds

(continued)

Table 6.2 (continued)

S. no.	Jurisdiction	Application no.	Title of invention	Details of patentable subject matter objection
12		AU2011225828 (also filed in the US and the EU but no PSM-based objection received)	Process, tube and device for the preparation of wound healant composition	Independent claim 10 and dependent claims 11, 13, and 15 do not meet the manner of manufacture requirement of the Patents Act because all that is defined is a collocation of components wherein all of the components are known
13		AU2014355213 (also filed in the US and the EU but no PSM-based objection received)	Novel ultrashort hydrophobic peptides that self-assemble into nanofibrous hydrogels and their uses	Claims 43–45 define a kit or pack of known integers and therefore do not define a 'patentable combination.'

The examiner applied the two-step evaluation procedure developed by the USPTO based on the *Alice/Mayo* test.[15] Step 1 requires determination of whether the claims are directed to a process, machine, manufacture, or a composition of matter. Step 2 analysis is a two-part analysis, with the first part, Step 2A requiring determination of whether the claims are directed towards a judicial exception, which includes laws of nature, natural phenomena (products of nature), or an abstract idea. Step 2B requires determination of whether the provided claim includes additional elements that add significantly more than the recited judicial exception. Considering Step 1 of the test, the examiner was satisfied that the present claim satisfies the requirement as the claimed subject matter is directed to a composition of matter. The results of the Step 2A analysis indicate that the subject matter claimed was directed to a judicial exception, namely naturally occurring cells. In light of this, it was necessary for the examiner to consider Step 2B.

The outcome of Step 2B analysis was that the provided claims did not recite additional elements that were seen by the examiner as being sufficient to amount to significantly more than the judicial exception, as pancreatic stellate cells, endothelial cells, and tumour tissue are all naturally occurring cells. The applicant was provided with a specific period to amend the application in light of the examiner's objections and to submit the amendments in response to the examiner's report. However, the patent applicant did not file any response to the examination report in the prescribed period of time and the application was later deemed abandoned.

The examination reports for the other US patent application that was objected to based on subject matter grounds related to a patent claim directed towards 'a tissue

[15] United States Patent and Trademarks Office (2019), *Manual of Patent Examining Procedure*, Part 2106; *Alice Corp. Pty. Ltd. v. CLS Bank Int'l*, 134 S. CT. 2347 [2014]; *Mayo Collaborative Services v. Prometheus Laboratories, Inc.*, 132 S.CT. 1289 [2012].

system, comprising: a support material, and a vascular network…'[16] The complete patent claim is as follows.

> A tissue system, comprising: a support material, and a vascular network comprising a plurality of channels disposed in the support material, wherein the vascular network is physiologically relevant[17]

Following examination, the patent examiner of USPTO responded that,

> … the claimed invention is directed to a judicial exception (i.e., a law of nature, a natural phenomenon, or an abstract idea) without significantly more. In this instant case, the claims were drawn to a natural product of vascularized tissue. Claim 1 is directed to a tissue product comprising a support material and a vascular network.[18]

Since none of the limitations recited in the claim amounted to significantly more than the judicial exception of a natural product, the claim was clearly considered to not be patentable subject matter.[19] The applicants subsequently amended the claim in order to overcome the 35 USC 101(of AIA)-based rejection. The amended claim reads as follows:

> A tissue system, comprising: an optically transparent support material; and a vascular network comprising a plurality of channels disposed in the support material, wherein the vascular network is physiologically relevant; and wherein at least some of the channels are structurally arranged according to an artificial predefined three-dimensional pattern.

By incorporating the phrase 'an optically transparent' before support material and adding 'wherein the vascular network is physiologically relevant; and wherein at least some of the channels are structurally arranged according to an artificial predefined three-dimensional pattern' the applicant transformed the claim into a claim for a non-natural product. In this way, the amended claim displayed 'markedly different characteristics' than naturally occurring vascularised tissue. The patent examiner considered the applicant's amendments and concluded that they had adequately addressed the objections. No additional subject matter-related objections were raised in the later examination reports.

EU

Patent applications with claims directed towards bioprinted tissues or organs filed with the EPO have received most objections (five patent applications) on patentable subject matter grounds in comparison with Australia and the US. Of a total of five applications that received subject matter objections, two applications were identified

---

[16] *Omnidirectional, multiaxial bioprinted tissue system, techniques and applications*, ch Lawrence Livermore National Security LLC, US20150050686A1 (2015) ('*Omnidirectional, multiaxial bioprinted tissue system, techniques and applications*').

[17] Ibid.

[18] Ibid.

[19] *Leahy-Smith America Invents Act 2011* (US) s 33(a).

for further examination of the detailed patent prosecution history. One of the EU patent applications that received objections is entitled 'Engineered three-dimensional connective tissue constructs and methods of making the same.' The objection was in respect of claims that included multipotent stem cells. The objected patent claims are as follows:

> Claim 1: An engineered, living, three-dimensional connective tissue construct comprising: connective tissue cells cohered to one another to provide a living, three-dimensional connective tissue construct; wherein the construct is substantially free of pre-formed scaffold at the time of use.
>
> Claim 3: The construct of claim 1, wherein the connective tissue cells comprise connective tissue cells derived in vitro from multi-potent cells.
>
> Claim 4: The construct of claim 3, wherein the multi-potent cells comprise one or more of: tissue-specific progenitors, mesenchymal stem/stromal cells, induced pluripotent stem cells, and embryonic stem cells.[20]

The EPO examiner examined these claims in light of Article 53(a) and Rule 28(c) of European Patent Convention (EPC), discussed above in Chap. 5.[21] Specifically, Article 53(a) requires patent examiners not to grant EU patents that claim,

> ... inventions the commercial exploitation of which would be contrary to 'ordre public' or morality; such exploitation shall not be deemed to be so contrary merely because it is prohibited by law or regulation in some or all of the Contracting States.[22]

Rule 28(1)(c) provides that,

> ...under Article 53(a), European patents shall not be granted in respect of biotechnological inventions which, in particular, concern the following: uses of human embryos for industrial or commercial purposes.[23]

In this case, the patent examiner raised the concern that, as totipotent cells, the claimed cells may come within this exception on the basis that they can differentiate into all somatic lineages.[24] These lineages include the germline and extra-embryonic tissues such as the placenta. Human totipotent stem cells are excluded from patentability under Rule 29(1) EPC as they represent a complete human body at the early stage of its formation.

Totipotent cells have the ability to produce all cell types in the body, including extra-embryonic (placental) cells. The only totipotent cells are embryonic cells within the first few cell divisions after fertilisation. Pluripotent cells, such as embryonic stem cells, may give rise to all of the cell types that make up the human body, but

---

[20] *Engineered three-dimensional connective tissue constructs and methods of making the same*, ch Organovo Inc, EP2861270A1 (2015) ('*Engineered three-dimensional connective tissue constructs and methods of making the same*').

[21] See Sect. 5.2.1 of Chap. 5.

[22] Ibid.

[23] *European Patent Convention*, opened for signature 5 October 1973, (entered in force 7 October 1977) Rule 28(1)(c).

[24] *European Patent Convention*, opened for signature 5 October 1973, (entered in force 7 October 1977) Article 53(a), Rule 28(1)(c).

cannot develop into an entire organism (or placenta) on their own. Multipotent cells may differentiate into many cell types but are more limited than pluripotent cells; multipotent cells include adult stem cells and cord blood stem cells.[25]

The applicant claimed multipotent stem cells as a part of an engineered, living, three-dimensional connective tissue construct. Even though the multipotent stem cells were not part of the first claim (the independent claim), they were added in the following dependent claims (claims 3–10) and hence the EU patent examiner raised an objection.[26] The inclusion of embryonic stem cells in one of the dependent claims was the main reason for the EU patent examiner's objection, and because embryonic stem cells can evolve into any form of cell, the objection was made under Article 53(a) and Rule (1)(c). The patent examiner also pointed out that the identified claims should be read while considering the disclosed specification and suggested deletion of the wording 'totipotent stem cells' or alternatively that the applicant submit amendments amending the term 'human totipotent cells' to 'non-human totipotent cells.' The patent applicant subsequently amended the claims. In later examination reports, the patent examiner raises no subject matter objections.

Another EU patent application entitled 'Method for 3d printing of vascularized tissues and organs' received an objection from the EPO based on patentable subject matter grounds.[27] The objected patent claim (Claim 9) was a dependent claim over Claim 1 which claimed,

> 3D printing process for the production of tissues and organs with vascular structures using
>
> (a) a droplet printer for photorealistic high resolution prints and
> (b) a device for applying electromagnetic waves comprising the steps:
>
> - Provision of at least one bio ink and one capillary ink with cells and cross-linking molecules in the droplet printer
> - Applying at least one drop of the bio ink and the capillary ink to a reaction plane
> - Bringing the electromagnetic waves into contact with the crosslinking molecules in these drops on the reaction level
>
> and activation of the crosslinking molecules by means of undirected or directed movements of the electromagnetic waves in the droplets, whereby crosslinked structures are formed and vascular structures are thus obtained.[28]

Following on from claim 1, the objected claim (claim 9) was directed towards,

> Method according to claims 1 to 8,
>
> characterized in that

[25] Stemcell.ny.gov. 2021. *What is the difference between totipotent, pluripotent, and multipotent? | NYSTEM*. [online] Available at: https://stemcell.ny.gov/faqs/what-difference-between-totipotent-pluripotent-and-multipotent [Accessed 30 November 2021].

[26] *European Patent Convention*, opened for signature 5 October 1973, (entered in force 7 October 1977) Article 53(a), Rule 28(1)(c).

[27] *Method for 3D printing of vascularized tissues and organs*, ch Technische Universitaet Berlin, EP3771547A1 (2021) ('*Method for 3D printing of vascularized tissues and organs*').

[28] Ibid.

> the reaction plane is positioned in a bioreactor which is connected to a blood circuit, in particular during or after the end of the pressure[29]
>
> On this occasion, the EPO patent examiner examined the above claims in light of Article 53 (c) of EPC (exceptions to patentability) and indicated that the subject matter of claim 9 related to a surgical or diagnostic method. The examiner further concluded that such procedures are excluded from patentability in light of Article 53 (c) and hence Claim 9 was not considered patentable.[30] Under EU law, although a 'method of treatment' is not patentable, a substance or composition may be considered patentable for a specific use in a method of surgery, therapy, or diagnosis.[31] This gives rise to the 'purpose limited product' claim, having the format '[substance or composition A] for use in [medical method B]'. Provided the method/process is based on the effect of a substance or composition, any method that would be objected to under Article 53 (c) EPC might theoretically be stated in the above-mentioned claim structure. The material or composition employed, or the method in which it is used, can provide novelty and inventive step in such a claim.[32] The same has been applied by the examiner in objecting the above provided claim.

6.3.3.2 Bioinks, Hydrogels, Biogels, Compositions, and Methods

Patent applications claiming bioinks, hydrogels, biogels, and relevant compositions as products and processes have been subjected to patentable subject matter objections in Australia and the EU. The patent prosecution data of US patent applications does not reveal any objections to this subject matter. Specifically, patent applications claiming hydrogels (one patent application) and biogels (two patent applications) were objected to in Australia whereas in the EU, patent applications claiming bioinks (one patent application) and hydrogels (two patent applications) were objected to.

Australia

Closer examination of the specific examination reports indicates that an Australian patent application with patent claims directed towards 'a kit for formation of biogel…' was objected to based on patentable subject matter grounds.[33] A total of 15 patent claims in this application (two independent claims and 13 dependent claims) were objected to by the patent examiner. The initial submitted version of the objected independent patent claims are as follows:

> Claim 1: A kit for formation of a biogel, which comprises: a plurality of carriers, each carrier having a plurality of fibrinogen binding moieties immobilised to the carrier; and, separately, fibrinogen, wherein each molecule of fibrinogen can bind at least two fibrinogen binding moieties.

---

[29] Ibid.

[30] *European Patent Convention*, opened for signature 5 October 1973, (entered in force 7 October 1977) Article 53(c).

[31] Ibid., Article 54(5).

[32] E. Asif, 'Exclusion of diagnostic, therapeutic and surgical methods from patentability' (2013) 18 *Journal of Intellectual Property Rights*.

[33] *Biogel*, ch Haemostatix Ltd, AU2007327100A1 (2008) ('*Biogel*').

> Claim 2: A kit for formation of a biogel, which comprises a plurality of carriers, each carrier having a plurality of fibrinogen binding precursors immobilised to the carrier, wherein each fibrinogen binding precursor can be converted to a 10 fibrinogen binding moiety; and fibrinogen, wherein each molecule of fibrinogen can bind at least two fibrinogen binding moieties.[34]

Considering the above claims, the patent examiner of IP Australia responded that,

> Claims 1-15 are directed at non-patentable collocations of known agents as they comprise known biomolecule immobilized carriers and known compounds (fibrinogen, coagulation factors) which can be stored separately and (as currently drafted) freely used for other purposes than the intended use of forming a biogel.

According to the patent examiner's comments, the applicant amended the objected claims to 'A kit when used for formation of a biogel, which comprises: a plurality of soluble carriers….' The applicant included the phrase 'when used' after the word 'kit' and added a word 'soluble' in the phrase 'a plurality of soluble carriers.' The patent examiner subsequently reversed the patentable subject matter objection.

According to Australian patent law, if an invention consists of a combination of known integers (as identified above), the invention will only be considered a manner of manufacture and be patentable if a claim identifies the integers as being in a physical arrangement that provides for a functioning interrelationship in ordinary usage.[35] Section 3.3 of Chap. 3 of this book has provided an in-depth doctrinal analysis of patentable subject matter-related Australian case law considering patent applications with product claims directed towards bioprinted tissues and bioinks.[36] In light of this, the original claims set out above only revealed the elements (biomolecule immobiliszed carriers, fibrinogen, and coagulation factors) in the form of a biogel with very little or no functioning interrelationship disclosed. This explains why the claims were considered by the patent examiner to fail the patentable subject matter requirement. In the present case, the applicant was given the opportunity to amend the objected claims to overcome the patentable subject matter objection and the amended claims were later approved by the patent examiner. The amended version of the claims are provided below.

> Claim 1 (Amended): A kit when used for formation of a biogel, which comprises: a plurality of soluble carriers, each carrier having a plurality of fibrinogen binding moieties immobilised to the carrier; and, separately, fibrinogen, wherein each molecule of fibrinogen can bind at least two of the fibrinogen binding moieties, such that when the carriers are contacted by the fibrinogen, a biogel is formed in which the fibrinogen molecules are linked together via the carriers by non-covalent bonds between the fibrinogen molecules and the fibrinogen binding moieties.
>
> Claim 2 (Amended): A kit when used for formation of a biogel, which comprises a plurality of soluble carriers, each carrier having a plurality of fibrinogen binding precursors immobilised to the carrier, wherein each fibrinogen binding precursor can be converted to a fibrinogen binding moiety; and fibrinogen, wherein each molecule of fibrinogen can bind at least two of

---

[34] Ibid.

[35] *British Celanese Ltd. v Courtaulds Ltd.* 52 RPC 171 at 193–194; IP Australia (2004), *Manual of Practice and Procedure,* November, Part 2.9.2.16.

[36] See Sect. 3.3 of Chap. 3.

> the fibrinogen binding moieties, such that when the carriers are contacted by the fibrinogen following conversion of the fibrinogen binding precursors to fibrinogen binding moieties, a biogel is formed in which the fibrinogen molecules are linked together via the carriers by non-covalent bonds between the fibrinogen molecules and the fibrinogen binding moieties.[37]

EU

Some of the patent applications with claims directed towards bioinks and hydrogel products and processes filed with EPO have also received objections on patentable subject matter grounds. Further analysis of the examination reports of specific patent applications reveals that an EU patent application with claims directed towards 'use of bioink' has received an objection from an EU patent examiner on such grounds. The objected patent claim (Claim 11) as initially submitted to the EPO was dependent on Claim 10 which claimed,

> Use of a RGD-modified nanocellulose and/or RGD-modified alginate as a bioink for 3D printing, with or without fibrin and/or human cells, preferably fibroblasts, to form a dermis-like construct or to produce tissue.[38]

Followed by the above, the objected claim (Claim 11) was directed towards,

> The use of claim 10 for drug discovery, and/or testing of cosmetics, and/or as a disease model, and/or for use in reconstructive surgery and tissue repair.[39]

The patent examiner responded that claim 11 partly related to methods for treatment of the human or animal body by surgery, excepted from patentability under Article 53(c) EPC.[40] In the present application, the patent examiner concluded that claim 11 was directed towards the use of bioink for the testing of cosmetics, and/or as a disease model, and/or for use in reconstructive surgery and tissue repair. In light of Article 53(c) of the EPC, the use of bioink in reconstructive surgery and tissue repair would fall under 'methods for treatment of the human or animal body by surgery' and hence would not be patentable.[41]

6.3.3.3 Comparative Analysis

A comparative analysis of the patent applications that were commonly filed in Australia, the US, and the EU indicates that the patent applications that received

---

[37] *Biogel*, ch Haemostatix Ltd, AU2007327100B2 (2014) ('*Biogel*').

[38] *Preparation and applications of rgd conjugated polysaccharide bioinks with or without fibrin for 3D bioprinting of human skin with novel printing head for use as model for testing cosmetics and for transplantation*, ch Cellink AB, EP3463822A1 (2019) ('*Preparation and applications of rgd conjugated polysaccharide bioinks with or without fibrin for 3D bioprinting of human skin with novel printing head for use as model for testing cosmetics and for transplantation*').

[39] Ibid.

[40] *European Patent Convention*, opened for signature 5 October 1973, (entered in force 7 October 1977) Article 53(c).

[41] See Sect. 5.4.3 of Chap. 5.

patentable subject matter objections in Australia did not receive any patentable subject matter-related objections in the US and the EU. For instance, a patent application with claims directed towards biogels received a patentable subject matter objection in Australia but not in the US and the EU.[42] However, further analysis of the first claim of this patent application filed in the US, the EU, and Australia reveals that the claims were actually amended before filing in the US and the EU patent offices. Specifically, the first claim of the patent application filed in Australia reads as follows.

> A kit for formation of a biogel, which comprises: a plurality of carriers, each carrier having a plurality of fibrinogen binding moieties immobilised to the carrier; and, separately, fibrinogen, wherein each molecule of fibrinogen can bind at least two fibrinogen binding moieties.[43]

The first claim of the same application filed in the EU reads as follows.

> A kit for formation of a biogel, which comprises:
>
> a) a plurality of carriers, each carrier having a plurality of fibrinogen binding moieties immobilised to the carrier; and, separately, fibrinogen, wherein each molecule of fibrinogen can bind at least two fibrinogen binding moieties; or
> b) a plurality of carriers, each carrier having a plurality of fibrinogen binding precursors immobilised to the carrier, wherein each fibrinogen binding precursor can be converted to a fibrinogen binding moiety; and fibrinogen, wherein each molecule of fibrinogen can bind at least two fibrinogen binding moieties.[44]

The applicant's first claim of the same application filed in the US reads as follows.

> A kit for formation of a biogel, which comprises: a plurality of carriers, each carrier having a plurality of fibrinogen binding moieties immobilised to the carrier; and, separately, fibrinogen, wherein each molecule of fibrinogen can bind at least two fibrinogen binding moieties.[45]

Although there was no change in the initial version of the first claim filed in the US, the applicant cancelled claims 1–15 before the start of the examination procedure in that jurisdiction. The same patent application filed in the EU, with the version of the first claim specified above, did not receive any patentable subject matter-based objections and was approved for grant. However, as discussed above, the first claim of the same patent application filed in Australia did receive a patentable subject matter-based objection, and this specific objection was withdrawn as the applicant then made minor amendments to the first claim in light of the subject matter-based objections. Considering this example, then it can be inferred that subject matter-based objections can be addressed with minor amendments in most cases in the US, the EU, and Australia. As seen above, patent applicants who are interested in filing the same patent application in multiple jurisdictions in most cases amend the claims

[42] *Biogel*, ch Haemostatix Ltd, AU2007327100A1 (2008) ('*Biogel*').

[43] Ibid.

[44] *Biogel*, ch Haemostatix Ltd, EP3095468A1 (2016) ('*Biogel*').

[45] *Biogel*, ch Haemostatix Ltd, US20100249044A1 (2010) ('*Biogel*').

according to jurisdiction-specific subject matter or patentability requirements to ease the process of patent approval.

In comparison with the objections based on other patentability criteria, the objections based on patentable subject matter are few in number in all the three jurisdictions. The further sections of this chapter will provide insights on objections based on other patentability criteria. Figure 6.3 provides an overview of the types of objections received in respect of patent applications filed in the US, EU, and Australia. Patent offices around the world have different patentability requirements in which novelty, obviousness, and utility or industrial applicability are most common. Patent prosecution data analysis of patent applications with claims directed towards bioinks and bioprinted tissues and organs did not reveal any objections related to utility or industrial applicability. Hence, no visualisation related to utility-based objections is reflected in Fig. 6.3. The absence of utility-based objections may be due to the fact that patent applications with claims directed towards bioinks and bioprinted tissues and organs would generally be considered useful for different applications such as tissue regeneration and organ transplantation.

6.3.4 Novelty

Patent systems around the globe include novelty or newness as one of the patentability criteria to grant a patent application.[46] Novelty of an invention is assessed in comparison with the prior art base as it existed before the priority of the claimed invention.[47] In comparison with patentable subject matter, assessment of the novelty of a claim is fact-based, as it compares the technology that is in existence prior to the disclosure of the claimed invention. Patent law in Australia has provisions on assessing inventions based on novelty criteria. According to Section 18(1)(b)(i) of the *Patents Act 1990*,

> …an invention is a patentable invention for the purposes of a standard patent if the invention, so far as claimed in any claim… when compared with the prior art base as it existed before the priority date of that claim…is novel[48]

Patent law in the US also has similar provisions for assessing novelty of an invention filed with the USPTO. Section 102 provides that,

> A person shall be entitled to a patent unless—the claimed invention was patented, described in a printed publication, or in public use, on sale, or otherwise available to the public before the effective filing date of the claimed invention.[49]

Article 54 of the EPC also provides similar criteria to assess novelty of patent applications filed with EPO, as provided by Article 54,

---

[46] Edward C. Walterscheid, 'Novelty in Historical Perspective (Part I)' (1993) 75 *Journal of the Patent and Trademark Office Society* 689.

[47] Ibid.

[48] *Patents Act 1990* (Cth) s 18(1)(b)(i).

[49] *Leahy-Smith America Invents Act 2011* (US) s 102.

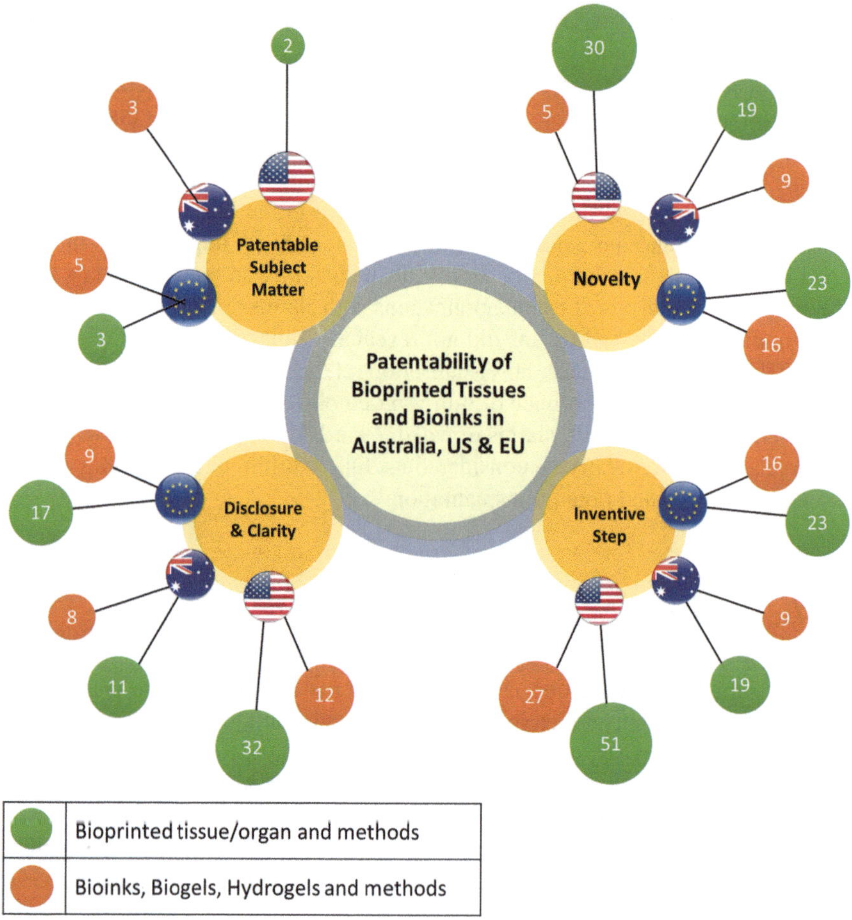

Fig. 6.3 A comparative network visualisation of types of objections received by patent applications claiming bioprinted tissues/organs, bioinks, and related sub-technologies in Australia, the US, and the EU

> An invention shall be considered to be new if it does not form part of the state of the art…The state of the art shall be held to comprise everything made available to the public by means of a written or oral description, by use, or in any other way, before the date of filing of the European patent application.[50]

Analysis of patent prosecution data provided insights on objections received by patent applications claiming bioprinted tissues, bioinks, and other relevant technologies based on the above novelty requirements of Australia, the US, and the EU. Patent applications with claims directed towards bioprinted tissues or organs and methods have received the highest number of novelty-based objections in each of

[50] *European Patent Convention*, opened for signature 5 October 1973, (entered in force 7 October 1977) Article 54.

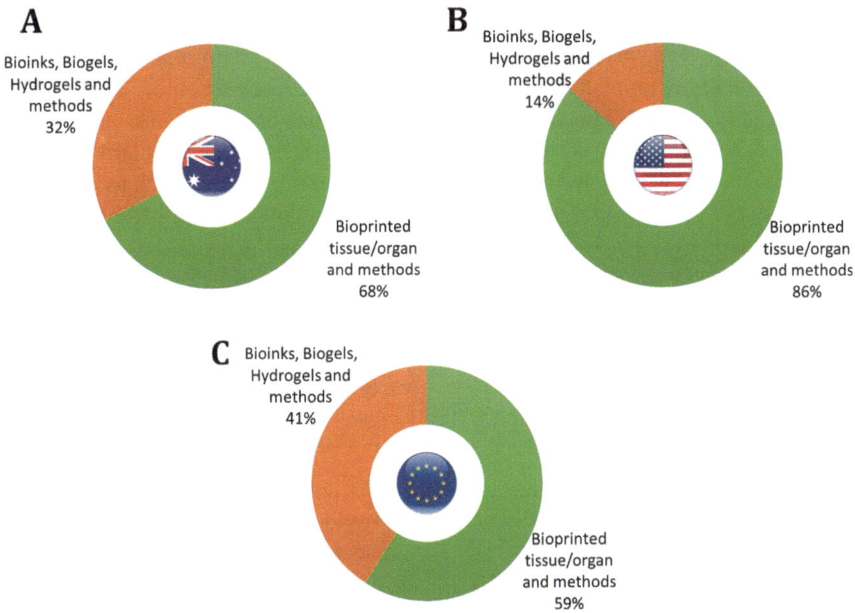

Fig. 6.4 A comparative visualisation of types of patent claims related to bioprinting that received objections based on novelty criteria of Australia, the US, and the EU

the three jurisdictions. The US has the highest percentage (86% of patent applications) of novelty-based objections followed by Australia with 68% and the EU with 59%. Patent applications with claims directed towards bioinks, biogels, hydrogels, and methods have also received novelty-based objections. The EU has the highest percentage (41%) of novelty-based objections on claims directed towards bioinks, biogels, hydrogels, and methods followed by Australia with 32% and the US with 14%. As provided in Fig. 6.3, a greater number of patent applications with claims directed towards bioprinted tissues or organs and methods received novelty-based objections in all the three jurisdictions in comparison with the number of patent applications with claims directed towards bioinks, biogels, hydrogels, and methods. This is also reflected in the percentage share shown in Fig. 6.4.

6.3.5 *Obviousness or Inventive Step*

Inventive step is one of the patentability criteria that must be satisfied for a patent application to be granted. Inventive step is also known as obviousness or inventiveness in different jurisdictions. It enables patent examiners to object to a patent application if the disclosed invention is obvious for a 'person having ordinary skill in the art' based on prior art.[51] Inventive step is a critical element of the patentability criteria based on which most of the patent applications are objected to, and this can also be seen in Fig. 6.3, where the number of patent applications that received objections based on inventive step are higher in comparison with objections based on other patentability criteria such as patentable subject matter, novelty, disclosure, and clarity. Even if some patent applications overcome the other patentability criteria such as patentable subject matter and novelty, they may receive objections based on inventive step.[52] Inventive step criteria are defined in a variety of ways in the patent laws of different jurisdictions; however, the meaning or the general idea of obviousness is directed to the same concept. Specifically, Section 18(1)(b)(ii) of the *Patents Act 1990* of Australia defines inventive step as follows:

> ...an invention is a patentable invention for the purposes of a standard patent if the invention, so far as claimed in any claim... when compared with the prior art base as it existed before the priority date of that claim...involves an inventive step.[53]

In the US, the inventive step requirement is known as the 'non-obviousness' condition and Section 103 of US Patent law provides conditions of patentability in light of non-obvious subject matter; according to which,

> a patent for a claimed invention may not be obtained [...] if the differences between the claimed invention and the prior art are such that the claimed invention as a whole would have been obvious before the effective filing date of the claimed invention to a person having ordinary skill in the art to which the claimed invention pertains.[54]

Article 52 of the EPC requires an invention to 'involve an inventive step,' and Article 56 further describes that 'an invention shall be considered as involving an inventive step if, having regard to the state of the art, it is not obvious to a person skilled in the art.'[55]

Analysis of patent prosecution data in light of the inventive step criteria in Australia, the US, and the EU provided insights on inventive step-based objections on patent applications with claims directed towards bioprinted tissues and bioinks. As visualised in Fig. 6.5, in comparison with objections received for bioinks, biogels, hydrogels, and methods, patent applications with product or process claims directed towards bioprinted tissue or organs have received more inventive step or obviousness-based objections in each the three jurisdictions, with Australia being the highest with 68% followed by the US with 65% and the EU with 59%. Secondly, product and process claims directed towards bioinks, biogels, and hydrogels have also received inventive step or obviousness-based objections with the EU having the

---

[51] Jeanne C. Fromer, 'The Layers of Obviousness in Patent Law' (2008) 22 *Harvard Journal of Law & Technology* 75.

[52] Lee Petherbridge and R. Polk Wagner, 'The Federal Circuit and Patentability: An Empirical Assessment of the Law of Obviousness' (2006) 85 *Texas Law Review* 2051.

[53] *Patents Act 1990* (Cth) s 18(1)(b)(ii).

[54] *Leahy-Smith America Invents Act 2011* (US) s 103.

[55] *European Patent Convention*, opened for signature 5 October 1973, (entered in force 7 October 1977) Articles 52, 56.

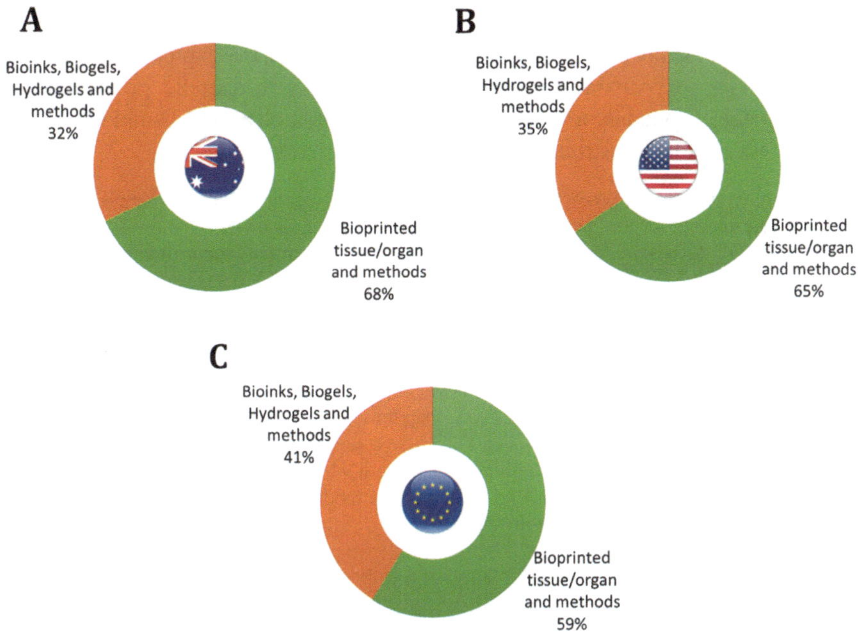

Fig. 6.5 A comparative visualisation of types of patent claims related to bioprinting that received objections based on inventive step criteria of **A** Australia, **B** United States, and **C** European Union

highest percentage at 41% followed by the US with 35% and Australia with 32%. As provided in Fig. 6.3, the number of patent applications with claims directed towards bioprinted tissues or organs and methods that received obviousness or inventive step-based objections are higher in all three jurisdictions in comparison with the number of patent applications with claims directed towards bioinks, biogels, hydrogels, and methods, and this is reflected in the percentage share that is shown in Fig. 6.5.

6.3.6 Disclosure and Clarity

Apart from the patentability criteria discussed above, patent applications must satisfy disclosure requirements. The written description of the invention provided in the patent application must disclose the process and way of manufacturing, utilising, or performing the invention in a complete, clear, succinct, and sufficient manner such that the person skilled in the relevant field can reproduce the invention. The disclosure and clarity requirements are important as the patent protection is only valid for a limited period (20 years from the priority date in most jurisdictions), and once the patent protection period is complete, the disclosed invention is open for public use. Hence, patent offices require patent applicants to disclose the invention in a clear and complete manner so that the disclosed invention can be easily practised or reproduced by a person skilled in the relevant art upon the expiry of the patent.[56]

Patent laws of Australia, the US, and the EU have specific provisions that require patent applicants to disclose an invention in a clear and complete manner. Section 40(2) of the Australian *Patents Act 1990* provides,

> a complete specification must: disclose the invention in a manner which is clear enough and complete enough for the invention to be performed by a person skilled in the relevant art; and disclose the best method known to the applicant of performing the invention.[57]

Similarly, Section 112 of US Patent law sets forth three basic requirements when filing a patent: Written Description, Enablement, and Best Mode. The 35 U.S.C Section 112(a) reads:

> The specification shall contain a written description of the invention, and of the manner and process of making and using it, in such full, clear, concise, and exact terms as to enable any person skilled in the art to which it pertains, or with which it is most nearly connected, to make and use the same, and shall set forth the best mode contemplated by the inventor or joint inventor of carrying out the invention.[58]

The EPC also provide similar requirements that need to be satisfied to grant an EU patent application. Specifically, the disclosure and clarity requirements are provided in Articles 83 and 84 of the EPC. Concerning sufficiency of disclosure, the EPC Article 83 provides that, 'The European patent application shall disclose the invention in a manner sufficiently clear and complete for it to be carried out by a person skilled in the art.'[59] Similarly, EPC Article 84 provides that, 'The claims shall define the matter for which protection is sought. They shall be clear and concise and be supported by the description.'[60]

Patent applications with claims directed towards bioprinted tissues or organs received the highest number of disclosure and clarity-based objections in each of the three jurisdictions as indicated in Fig. 6.3 and further observed in the percentage share shown in Fig. 6.6. Patent applications with claims directed towards bioinks, biogels, hydrogels, and methods also received disclosure and clarity-based objections. Australia has the highest percentage (42%) of disclosure and clarity-based objections on claims directed towards bioinks, biogels, hydrogels, and methods followed by the EU with 35% and the US at 27%.

---

[56] Edimilson Junqueira Braga et al., 'The role of specification in patent applications: A comparative study on sufficiency of disclosure' (2018) 53 *World Patent Information* 58–65.

[57] *Patents Act 1990* (Cth) s 40(2).

[58] *Leahy-Smith America Invents Act 2011* (US) s 112.

[59] *European Patent Convention*, opened for signature 5 October 1973, (entered in force 7 October 1977) Article 83.

[60] Ibid., Article 84.

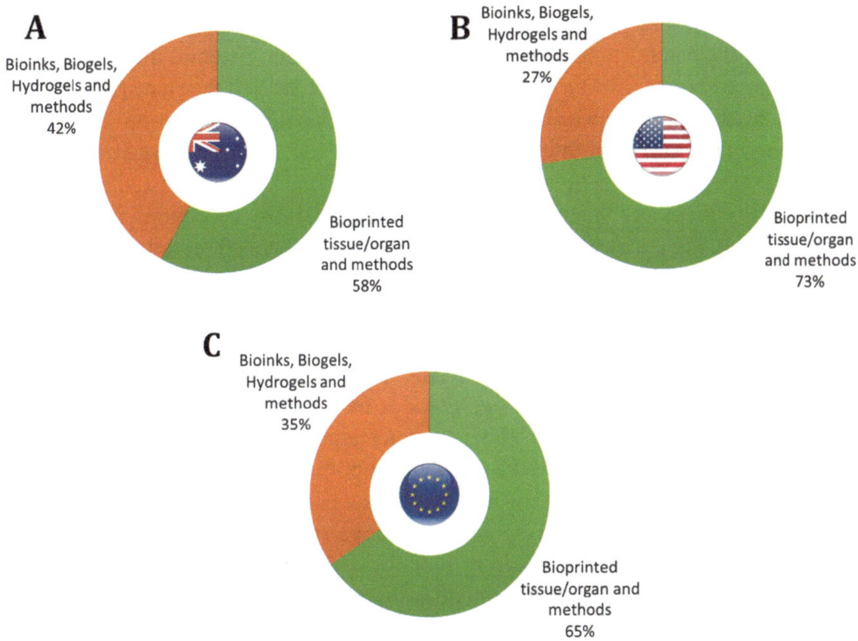

Fig. 6.6 A comparative visualisation of types of patent claims related to bioprinting that received objections based on disclosure and clarity requirements of **A** Australia, **B** United States, and **C** European Union

6.3.7 Comparative Patentability Analysis of Patent Family Members Commonly Filed in Australia, the US, and the EU

Patent family members are patent applications with similar subject matter that are filed in different jurisdictions. Patent families help patent applicants obtain protection for the same invention in multiple jurisdictions. A comparative patentability analysis of patent family members with claims directed towards bioprinted tissues and bioinks that were commonly filed in Australia, the US, and EU was conducted in respect of patentability objections (novelty, inventive step, and disclosure or clarity). Table 6.3 provides an overview of patentability objections in Australia, the US, and the EU for the patent family members having patent claims direct towards the same subject matter. This comparative analysis provides insights on how the patentability criteria of Australia, the US, and the EU are applied on claims with similar subject matter. A total of 30 patent family members (each family includes three patent applications with similar subject matter filled in all the three jurisdictions) were commonly filed in each jurisdiction but patent examination of 14 patent family members is still in progress in one or more of the jurisdictions. Hence, 16 patent family members for which at least the first examination reports are available were considered for this comparative analysis. Table 6.3 indicates that of a total of 16 patent family members, four patent family members received the same patentability outcome in all three jurisdictions. Eleven patent family members received the same patentability outcome in at least two jurisdictions. For example, a patent application with claims directed towards 'a living, three-dimensional tissue construct' received novelty and inventive step-based objections in all three jurisdictions.[61] In contrast, there are few patent family members that received different patentability objections for the same subject matter. For example, a patent application with claims directed towards 'a 3D tissue culture model comprising a drop of cells encapsulated in a hydrogel formed by a 3D bioprinter and the process of producing the same' received objections based on novelty and inventive step in Australia, inventive step and specification in the US, and only clarity-based objections in the EU.[62]

Generally, international patent applications having patent family members with claims directed towards similar subject matter and claiming priority of the same parent patent application are provided with an opportunity to amend the patent application or claims while entering the national phase. For example, an EU patent application which is a family member of the Australian patent application with claims directed towards 'a 3D tissue culture model…and process' was amended before examination. The amendment was made in light of the international search report and the examination reports of other jurisdictions and hence did not receive the same novelty and inventive step-based objections similar to other jurisdictions.[63] Considering the conclusions of the comparative analysis conducted in this chapter, it can also be inferred that even though there are some differences in the application of the disclosure and clarity requirements of each of the three jurisdictions, the application of the novelty and inventive step requirements seems to be very similar in these jurisdictions.

6.4 Discussion

The patent laws of Australia, the US, and the EU have a diverse history. Particularly in the EU, legislation has strictly circumscribed the limits on patentability, whereas in the US and Australia, the patent laws and the patentability requirements have evolved significantly through case law.[64] Although the general patentability criteria

[61] *Engineered tissues for in vitro research uses, arrays thereof, and methods of making the same*, Priv L No Australia, ch Organovo Inc, AU2012308715A1 (2014) ('*Engineered tissues for in vitro research uses, arrays thereof, and methods of making the same*').

[62] *Process for printing 3D tissue culture models*, Priv L No Australia, ch Inventia Life Science Pty Ltd, AU2016297675B2 (2020) ('*Process for printing 3D tissue culture models*').

[63] *Process for printing 3d tissue culture models*, ch Inventia Life Science Pty Ltd, EP3325611B1 (2020) ('*Process for printing 3d tissue culture models*').

[64] Minssen and Mimler (n; Olumayowa O Adesanya, 'Patenting bioprinted structures in a clime of moral uncertainty: time to amend the patents act?' (2019) 29(4) *Australian Intellectual Property Journal* 222–238; Laura C. Whitworth, 'Comparison of the Implementation of Statutory Patent

Table 6.3 A comparative overview of patent applications that are commonly filed in the US, EU, and Australia and the types of patentability objections raised by specific jurisdiction. The wording 'Bioprinted tissue/organ, M' provided in some of the columns indicate that specific patent application also included method claims related to bioprinted tissue or organ

S. no.	Patent application number	Title	Australia			US			EU		
			Novelty	Inventive step	Clarity	Novelty	Inventive step	Specification	Novelty	Inventive step	Clarity and conciseness
1	AU2007327100B2	Biogel	Biogel/ Bioink	Biogel/ Bioink			Biogel/ Bioink		Biogel/ Bioink	Biogel/ Bioink	Biogel/ Bioink
2	AU2012308715A1	Engineered tissues for in vitro research uses, arrays thereof, and methods of making the same	Bioprinted tissue/ organ	Bioprinted tissue/ organ		Bioprinted tissue/ organ	Bioprinted tissue/ organ		Bioprinted tissue/ organ	Bioprinted tissue/ organ	
3	AAU2012308724A1	Platform for engineered implantable tissues and organs and methods of making the same			Bioprinted tissue/ organ	Bioprinted tissue/ organ	Bioprinted tissue/ organ	Bioprinted tissue/organ	Bioprinted tissue/ organ	Bioprinted tissue/ organ	Bioprinted tissue/ organ
4	AU2011318437B2	Devices, systems, and methods for the fabrication of tissue	Bioprinted tissue/ organ, M	Bioprinted tissue/ organ, M	Bioprinted tissue/ organ, M		Bioprinted tissue/ organ, M	Bioprinted tissue/organ, M	Bioprinted tissue/ organ, M		Bioprinted tissue/ organ, M

(continued)

Table 6.3 (continued)

S. no.	Patent application number	Title	Australia			US			EU		
			Novelty	Inventive step	Clarity	Novelty	Inventive step	Specification	Novelty	Inventive step	Clarity and conciseness
5	AU2016203930A1	Engineered tissues for in vitro research uses, arrays thereof, and methods of making the same	Bioprinted tissue/ organ	Bioprinted tissue/ organ		Bioprinted tissue/ organ	Bioprinted tissue/ organ		Bioprinted tissue/ organ	Bioprinted tissue/ organ	
6	AU2013277275B2	Engineered three-dimensional connective tissue constructs and methods of making the same	Bioprinted tissue/ organ	Bioprinted tissue/ organ		Bioprinted tissue/ organ	Bioprinted tissue/ organ		Bioprinted tissue/ organ	Bioprinted tissue/ organ	
7	AU2014389440A1	Engineered three-dimensional breast tissue, adipose tissue, and tumor tissue model	Bioprinted tissue/ organ, M	Bioprinted tissue/ organ, M					Bioprinted tissue/ organ, M	Bioprinted tissue/ organ, M	
8	AU2017200162A1	Engineered three-dimensional connective tissue constructs and methods of making the same	Bioprinted tissue/ organ, M	Bioprinted tissue/ organ, M		Bioprinted tissue/ organ, M	Bioprinted tissue/ organ, M	Bioprinted tissue/organ, M	Bioprinted tissue/ organ, M	Bioprinted tissue/ organ, M	Bioprinted tissue/ organ, M

(continued)

Table 6.3 (continued)

S. no.	Patent application number	Title	Australia			US			EU		
			Novelty	Inventive step	Clarity	Novelty	Inventive step	Specification	Novelty	Inventive step	Clarity and conciseness
9	AU2015328173A1	Engineered renal tissues, arrays thereof, and methods of making the same	Bioprinted tissue/ organ, M	Bioprinted tissue/ organ, M	Bioprinted tissue/ organ, M	Bioprinted tissue/ organ, M	Bioprinted tissue/ organ, M		Bioprinted tissue/ organ, M		Bioprinted tissue/ organ, M
10	AU2013249569B2	Devices, systems, and methods for the fabrication of tissue utilizing UV cross-linking	Bioink	Bioink			Bioink	Bioink	Bioink	Bioink	
11	AU2017202721A1	Platform for engineered implantable tissues and organs and methods of making the same	Bioprinted tissue/ organ	Bioprinted tissue/ organ		Bioprinted tissue/ organ	Bioprinted tissue/ organ	Bioprinted tissue/organ	Bioprinted tissue/ organ	Bioprinted tissue/ organ	Bioprinted tissue/ organ
12	AU2015352873A1	Improved methods for tissue fabrication	Bioprinted tissue/ organ, M	Bioprinted tissue/ organ, M			Bioprinted tissue/ organ, M	Bioprinted tissue/organ, M	Bioprinted tissue/ organ, M	Bioprinted tissue/ organ, M	

(continued)

Table 6.3 (continued)

S. no.	Patent application number	Title	Australia			US			EU		
			Novelty	Inventive step	Clarity	Novelty	Inventive step	Specification	Novelty	Inventive step	Clarity and conciseness
13	AU2016289944B2	Vascularized in vitro perfusion devices, methods of fabricating, and applications thereof	Bioprinted tissue/ organ, M	Bioprinted tissue/ organ, M	Bioprinted tissue/ organ, M	Bioprinted tissue/ organ, M	Bioprinted tissue/ organ, M	Bioprinted tissue/organ, M	Bioprinted tissue/ organ, M	Bioprinted tissue/ organ, M	Bioprinted tissue/ organ, M
14	AU2016297675B2	Process for printing 3D tissue culture models	Bioprinted tissue/ organ, M	Bioprinted tissue/ organ, M			Bioprinted tissue/ organ, M	Bioprinted tissue/organ, M			Bioprinted tissue/ organ, M
15	AU2014296246B2	Automated device, systems, and methods for the fabrication of tissue	Bioink	Bioink	Bioink		Bioink	Bioink		Bioink	Bioink
16	AU2014277667	Differentiation of pluripotent stem cells to form renal organoids	Bioprinted tissue/ organ	Bioprinted tissue/ organ	Bioprinted tissue/ organ				Bioprinted tissue/ organ	Bioprinted tissue/ organ	Bioprinted tissue/ organ

of these three jurisdictions mostly follow a similar pattern due to the requirements associated with TRIPS compliance, the earlier chapters of this book illustrate that the requirements of patentable subject matter differ to some extent. As such, when inventions related to advanced technologies such as 3D bioprinting are examined in light of these diverse subject matter requirements, the patent eligibility outcomes may not always be identical.[65]

The legal status analysis provided in Fig. 6.1 also reflects differences in the patent application numbers that are granted and active in the three jurisdictions. Similarly, the patent claim type analysis (provided in Fig. 6.2) indicates that patent applications with claims directed towards products and processes of bioprinting and bioinks are fewer (15%) in Australia in comparison with the US (58%) and the EU (61%). In contrast, patent claims directed towards only processes are higher (55%) in Australia compared to the US (17%) and the EU (7%). One of the reasons that could have given rise to this scenario would be that the patent claims directed towards processes in Australia could be drafted to claim product-by-process.[66] Product-by-process claims define a product in the form of a process.[67] Such claims are used when the product's manufacturing process steps are intended to impose distinctive structural characteristics on the product, or where the product can only be defined by the process steps by which it is manufactured.[68]

The patentable subject matter-based objections relating to product and process claims of bioprinted tissues or organs in the US are more focused on examining the claims considering the judicial exceptions based on 'products of nature' and the inclusion of 'additional elements' that are sufficient to amount to significantly more than the judicial exception and would make a claim 'non-naturally occurring.' However, the patentable subject matter-based objections related to product and process claims of bioprinted tissues or organs in the EU are more focused on whether the recited claim claims subject matter related to human totipotent stem cells as they represent a complete human body at the early stage of its formation. The number of patentable subject matter objections in the EU are higher in comparison with the patentable subject matter objections of the US and Australia as European patent law has specific patentable subject matter restrictions different to those of the US and Australia. Specifically, Article 53(c) of EPC excludes patent applications with claims directed towards methods for treatment of the human or animal body by surgery and Article 53(a) and Rule 28(c) of EPC exclude human totipotent cells from patentability. As discussed in Sect. 6.3.3.1 of this chapter, once a patent claim has been objected to be based on patentable subject matter grounds, the patent examiner

---

Eligibility Requirements Applied to Gene Patents in the European Union, the United States, and Australia' (2016) 56 *IDEA* 449.

[65] Tabrez Y. Ebrahim, '3D Bioprinting Patentable Subject Matter Boundaries' (2017) 41 *Seattle University Law Review* 1.

[66] IP Australia (2004), *Manual of Practice and Procedure,* November, Part 2.11.2.3.11.

[67] Ibid.

[68] Eric P. Mirabel, 'Product-by-process claims: a practical perspective' (1986) 68 *Journal of the Patent and Trademark Office Society* 3.

provides details on specific wording or phrases related to patent-ineligible subject matter used in the claim. Patent applicants are provided with a specific time within which they may amend the objected claims. The patent applicant may then amend the claims to address these objections, and, in most cases, the amendments constitute minor tweaks to the claims, including deletion of objected wording or phases or replacing such wording or phases to ensure the claim reads as an 'artificially created state of affairs.' In certain cases, such amendments are sufficient to overcome objections based on patentable subject matter grounds. This is also evident in the example of a US patent discussed in Sect. 6.3.3.1 of this chapter.

According to the patentable subject matter requirements of Australia as discussed in Sect. 3.3.1 of Chap. 3, an invention will only be considered patentable if it satisfies the requirements of manner of manufacture.[69] It is particularly important to note that patent claims directed towards bioprinted tissues and bioinks must be drafted in a way that the elements (such as stem cells, growth factors, bioactive molecules, and the biomaterial) claimed are in a functional relationship to each other so that the claim as a whole would satisfy the manner of manufacture requirements. Although there are few subject matter objections in Australia, it appears that this may be because applicants draft their claims in such a way as to avoid objections. The post-objection claim amendment analysis in Sect. 6.3.3.2 of this chapter illustrates that claims can readily be amended to avoid subject matter objections.

The US perspective of analysing the patent claims directed towards bioprinted tissues in light of Section 101 is somewhat different from Australia.[70] Recent patent prosecution research conducted by researchers from Lexis Nexis using US patent prosecution data related to life sciences and high-tech sectors provided examination trends related to patentable subject matter (Section 101) and disclosure and claim language (Section 112).[71] The research conducted by Lexis Nexis indicated that during the period 2015–2020, an average of 10% (as a percentage of total rejections) of patent applications related to Biology and Organic Chemistry received rejections based on patentable subject matter requirements (Section 101). Similarly, an average of 45–50% (in a percentage of total rejections) of patent applications related to Biology and Organic chemistry received rejections based on disclosure and claim language requirements (Section 112). When we compare these numbers with the results of the patent prosecution analysis conducted in this chapter, the ratio of patentable subject matter-based objections almost reflects the same trend. Specifically, in the present patent prosecution analysis, the total patent applications with patentable subject matter-related objections number 13. On the other hand, total patent applications with novelty, inventive step, and disclosure-related objections are 102, 145, and 89, respectively.

---

[69] *Patents Act 1990* (Cth) s 18(1)(a); See Sect. 3.3.1 of Chap. 3.

[70] *Leahy-Smith America Invents Act 2011* (US) s 101.

[71] Quinn, Gene et al., "LexisNexis IP—Trends in Patent Prosecution—Nov 30, 2021—IPWatchdog.Com | Patents & Intellectual Property Law," *IPWatchdog.com | Patents & Intellectual Property Law* (November 14, 2001) https://ipwatchdog.com/videos/lexisnexis-ip-trends-patent-prosecution-nov-30-2021/.

In the US, the judicial exceptions play a crucial role in assessing patentable subject matter requirements and specifically the *Alice/Mayo* steps are required to be satisfied for patent claims directed towards bioprinted tissues.[72] As such, US patent applications with claims directed towards bioprinted tissues or bioinks must be drafted with a special focus on including additional elements sufficient to amount to 'significantly more' than the judicial exception in order to pass the *Alice/Mayo* steps.[73] Similar to the process in Australia and the EU, the patent claims that received patentable subject matter-based objections in the US may be amended to overcome the objections.

The analysis presented in this chapter shows that patentable subject matter requirements in the EU are more focused towards exclusion of stem cells that are capable of becoming a complete human body at the early stage of its formation.[74] The present analysis also indicate that stem cells that are unable to develop into an embryo or an early stage of the human body are used to develop inventions with the claims directed towards bioprinted tissues and bioinks. On the other hand, patent claims directed towards methods of use of bioprinted tissues or bioinks are recommended to be drafted so as to not incorporate their use in surgery, as such claims would fall under 'methods for treatment of the human or animal body by surgery' and may not be considered patentable under EU law.[75] The patentable subject matter requirements applied by the EPO are more focused on Article 53 and Rule 28 that enables EU patent examiners to object to claims with subject matter directed towards 'use of human embryos for industrial or commercial purposes.'[76] However, the patentable subject matter requirements of the US are more focused on the 'products of nature' doctrine and the absence of additional elements that are sufficient to amount to significantly more than the judicial exception.[77] On the other hand, the patentable subject matter requirements in Australia are more focused on the manner of manufacture requirement that restricts claims directed towards known elements that is similar to the 'products of nature' requirement in the US.[78] Although the requirements for patentable subject matter in Australia and the US appear to be similar, the application and interpretation of these requirements varies.

According to the ratio of patentable subject matter-based objections in light of total patent applications analysed in this study and the claim analysis conducted in Sect. 6.3.3 of this chapter, it is evident that most of the patent applications claiming bioprinted tissues and bioinks were drafted while keeping in mind the jurisdiction-specific patentable subject matter requirements. Additionally, the applicants were also provided with a chance to amend the patent claims during the prosecution stage to exclude subject matter that was not patent eligible. It is evident that an initial

[72] See Sect. 4.5 of Chap. 4.

[73] See Sect. 4.4 of Chap. 4.

[74] See Sect. 5.4.2.2 of Chap. 5.

[75] See Sect. 5.4.3 of Chap. 5.

[76] See Sect. 5.4.2.1 of Chap. 5.

[77] See Sect. 4.4 of Chap. 4.

[78] See Sect. 3.3.1 of Chap. 3.

patentable subject matter objection is not a final decision on a patent claim or a patent application.

While the present study focuses primarily on analysing patentable subject matter-based objections in Australia, the US, and the EU, an extended analysis was performed to analyse how patent examiners are applying other types of patentability criteria such as novelty, obviousness, and disclosure requirements to raise objections. This extended patent prosecution data analysis has provided insights on the types of objections received by patent applications claiming bioprinted tissues and bioinks. The insights of the extended patent prosecution data analysis indicates that most of the patent applications claiming bioprinted tissues and bioinks as products and processes were objected to be based on the novelty and inventive step criteria. Considering these results, it can be inferred that some of the patent applications with claims directed towards bioprinted tissues and bioinks that satisfied the patentability subject matter requirements were unable to overcome the novelty or inventive step requirements in all three jurisdictions.

A further comparative analysis of individual patent applications that received novelty and inventive step-based objections indicate that most of the patent applications that received these objections in one jurisdiction had also received novelty and inventive step-based objections in the other jurisdictions. For example, patent applications that are listed in Table 6.3 indicate that 10 out of 16 patent applications received inventive step objections in all three jurisdictions. This reflects the fact that, even though patentable subject matter requirements and their implications in the jurisdictions considered are comparatively different, the applications of novelty and inventive step requirements on patent applications with claims directed towards bioprinted tissues and bioinks are relatively analogous. This may be because patentability criteria such as novelty and inventive step are more fact based.

As the patent prosecution data of these jurisdictions is publicly available and accessible, the novelty and inventive step objections (based on relevant prior art) raised by one patent office (for example, Australia) are also accessible to the other patent offices. For example, a patent application is commonly filed across these jurisdictions, and the patent examination process may be initiated in any one of the jurisdictions (for example, Australia). For Australian applications, if the Australian patent examiner raised an objection based on novelty criteria in the first examination report, this is accessible to patent examiners of other jurisdictions. In most cases, patent examiners cite the prior art that was cited by the patent examiner of another jurisdiction and then raise novelty and inventive step objections. According to the comparative analysis provided in Sect. 6.3.7 and Table 6.3 of this chapter, the application of novelty and inventive step requirements in the three judications follows a similar path. Considering the inventive step requirements of Australia, 15 of 16 patent family members filed in Australia received inventive step-based objections which is slightly higher than the US (14 of 16) and EU (13 of 16). Some subject experts argue that Australia's inventive step requirements are comparatively lower than that of the US.[79] However, comparative analysis data indicates that the application of inventive step criteria (by Australian patent examiners) on patent applications with claims directed towards bioprinted tissues and bioinks is stringent (reflecting the raising the bar amendments) in comparison with the application of inventive step requirements in the US and the EU.[80]

Disclosure requirements have become increasing important in the examination process and the results of this patent prosecution analysis reflect this.[81] As provided in Sect. 6.3.6 of this chapter, patent applications with claims directed towards bioprinted tissues/organs have received more disclosure and/or clarity-based objections in all three jurisdictions.[82] Figure 6.3 illustrates that patent applications from the US received the greatest number of disclosure-based objections (44 applications) followed by the EU and Australia. There are multiple reasons that generally lead to disclosure-based objection, and particularly in the US, such objection may be raised by the examiner if the patent specification is not in full, clear, concise, and exact terms that enable a person skilled in the art to practise the invention.[83] Australian and EU patent laws also have similar provisions that enable patent examiners to object to claims that lack clarity and conciseness. According to the details provided in Table 6.3, four patent applications have received disclosure or clarity-based objections across the three jurisdictions. For example, one patent application universally received disclosure or clarity-based objections in all jurisdictions due to lack of enablement.[84] Specifically, the EPO examiner indicated that,

> …the subject matter claimed comprises organs, including bladder, uterus, stomach etc. The teaching of the present application would not enable the skilled person to prepare such organs, having complicated structural features and requiring complex vascularization, innervation etc.

The examiner indicated that the application was not disclosed in a manner sufficiently clear and complete for it to be carried out by a person skilled in the art.[85] The

---

[79] Hazel V.J. Moir, 'Empirical evidence on the inventive step' (2013) *European Intellectual Property Review*; *Review of intellectual property legislation under the Competition Principles Agreement—Final Report* Canberra: Commonwealth of Australia, *IP Australia*, ('*Review of intellectual property legislation under the Competition Principles Agreement—Final Report*').

[80] Co Ld v Cocker Brothers Ld, 'Less than obvious: the possible implications of the Productivity Commission's proposed changes to inventive step in Australia' (2016); *Inventive step and raising the bar further?: A proposal to assist examiners*, 107) (Intellectual Property Forum: journal of the Intellectual and Industrial Property Society of Australia and New Zealand) annex 0815-2098 ('*Inventive step and raising the bar further?: A proposal to assist examiners*').

[81] Nefissa Chakroun, 'Vagueness of patent claim language, claim construction and patent infringement—what a mess!' (2020) 69(11) *GRUR International* 1097–1104.

[82] See Sect. 6.3.6 of this chapter.

[83] *Leahy-Smith America Invents Act 2011* (US) s 112.

[84] *Platform for engineered implantable tissues and organs and methods of making the same*, ch Organovo Inc., EP2755599B1 (2020) ('*Platform for engineered implantable tissues and organs and methods of making the same*').

[85] *European Patent Convention*, opened for signature 5 October 1973, (entered in force 7 October 1977) Article 83.

same application received an objection on similar grounds by the Australian examiner under Section 40, referring to the conclusions of EPO examiner. In light of this illustration and the disclosure criteria, the claimed and disclosed subject matter in the patent application must satisfy the disclosure and clarity requirements, regardless of the type and the level of complexity of the technology (such as bioprinted tissue/organ).

The extended analysis in light of patentability criteria other than patentable subject matter also indicates a trend whereby the ratio of objections related to patent applications with claims directed towards bioprinted tissues or organs and methods is higher than bioinks, biogels, hydrogels, and methods. This may be in part due to the fact that patent application filings related to bioprinted tissues or organs and methods are higher in number compared to patent application filings related to bioinks, biogels, hydrogels, and methods in each of the three jurisdictions.

6.5 Guidance to Patent Applicants Useful at Different Stages of Patent Prosecution

While the patentable subject matter requirement can be seen to be a potential hurdle for patenting specific bioprinting claims, these challenges can be overcome using different strategies. This section provides some modest suggestions and strategies to patent applicants to satisfy the patentable subject matter requirements in all the three jurisdictions. The suggestions provided in this section are applicable at different stages of the patent application life cycle that includes drafting, filing, and prosecution. While some of the below provided suggestions are currently practised at different prosecution stages, it is important to know how, when, and at what stage of patent prosecution a specific suggestion is applicable to bioprinting technologies.

6.5.1 Withdrawal or Abandonment of Patent Application

If a filed patent application includes claims directed towards non-patentable subject matter, it can be withdrawn or abandoned by the patent applicant. However, it is important to note that all three patent offices have specific provisions applicable to withdrawal or abandonment of patent applications.[86] If an applicant wishes to withdraw or abandon a patent application, the request to withdraw or abandon the application must be made before the publication (18 months from earliest priority

[86] *Patents Act 1990* (Cth) s 141; IP Australia (2004), *Manual of Practice and Procedure,* November, Part 2.24.2.1; *Leahy-Smith America Invents Act 2011* (US) s 1.138; United States Patent and Trademarks Office (2019), *Manual of Patent Examining Procedure*, Part 711; Directive 98/44/EC of the European Parliament and of the Council, on the legal protection of biotechnological inventions, *Parliamentary Debates*, 06 July 1998 (n Rule 15. *European Patent Office,* Guidelines for Examination, Withdrawal of application or designation, 8.1.

date in all the three jurisdictions) of the patent application in the official journal of the relevant IP office.[87] Once the patent application is published, the subject matter disclosed in the patent application would be considered to be a public disclosure and become part of the prior art, and hence, the withdrawal request may be refused at that stage.[88] For example, if a patent application is submitted by an applicant who is unaware of the requirements for patentability and claimed non-patentable subject matter, withdrawing or abandoning the application prior to the publication deadline would enable the applicant to review it again with the assistance of a patent attorney and submit an updated version of the application.

6.5.2 *Amendment of Patent Claims*

Section 6.3.3 of this chapter provided analysis of different types of product and process claims related to bioprinted tissues and bioinks that received objections from different patent offices.[89] Guidelines in all three jurisdictions make provision for applicants to amend the patent specification and claims at different stages of patent application and prosecution. The availability of amendment provisions and the conclusions of the analysis provided in this chapter support that patent claims identified to be claiming non-patentable subject matter can be amended to comply with the provisions of relevant patent offices.[90] For example, in Australia, if an applicant wishes to amend a patent application before examination, such an amendment would be considered a voluntary amendment which can only be made by paying applicable amendment fees.[91] However, amendments made after examination in light of the examiner's suggestions do not attract any amendment fees in Australia.[92] This section will now elaborate on how and when amendments are allowed to be made to patent claims.

6.5.2.1 Amendments to National Phase Application Following a PCT Application

A national phase patent application followed by a Patent Cooperation Treaty (PCT) or international patent application must be filed within 30 months from the earliest

---

[87] IP Australia. 2022. *Types of patents.* [online] Available at: https://www.ipaustralia.gov.au/patents/understanding-patents/types-patents [Accessed 14 August 2022].

[88] *Patents Act 1990* (Cth) s 53.

[89] See Sect. 6.3.3 of this chapter.

[90] Ibid.

[91] *Manual of Practice and Procedure,* November, Part 2.23.4.1.

[92] Ibid.

priority date in the member states.[93] The PCT enables international patent applicants to amend a patent specification including patent claims according to the provisions of jurisdiction-specific patent laws.[94] In doing so, the applicant must ensure any amendments are made within the scope of initially claimed subject matter and not introduce any new subject matter.[95] Despite some of the similarities, patent provisions in relation to patentable subject matter generally vary between jurisdictions, and accordingly, patent claims that are considered patentable in one jurisdiction may not be considered patentable in other jurisdictions. The conclusions in Chaps. 3, 4, and 5 reflect the similarities and differences in the subject matter provisions of Australia, the US, and Europe, and correspondingly, jurisdictional divergences in the patentability of bioprinted tissues and bioinks. To minimise subject matter-related objections, it is advised to amend a patent application, including the patent claims, in line with the provisions of the particular jurisdiction before entering the national phase.

6.5.2.2 Amendments Considering the Examination Reports of Family Members

By the time the national phase application is anticipated to be filed in one of the member states under the PCT pathway, the examination report of the patent application (with the same subject matter) originally filed in the first jurisdiction will be accessible. This initial examination report provides an opportunity for the patent applicant to amend the patent claims of the national phase application before filing. It is important to note that amendments at this stage are more relevant if there are similarities in the patent provisions in the jurisdictions in which applications have been filed. For example, a patent application with claims directed towards 'a surgical method of bioprinting a tissue or bioink' may not be considered patentable in Australia or Europe.[96] Assume that the applicant wants to submit a national phase application in Australia before 30 months have passed after the earliest priority date and that the same patent application was initially filed in Europe, followed by a PCT application. If before filing the Australian application, the patent applicant received the examination report from the European patent office detailing that the claims directed towards the claims are considered not patentable, this provides the applicant with an opportunity to amend specific patent claims before filing a national phase application in Australia to avoid any similar objections. In such a case, specific claims may be filed in an amended form as, for example, 'a method of bioprinting a tissue or bioink' excluding the elements of the patent claim that are directed towards surgery.

---

[93] *Patent Cooperation Treaty (PCT)*, opened for signature 19 June 1970, (entered into force 24 January 1978) Article 22.

[94] Ibid., Article 28.

[95] Ibid.

[96] See Sect. 5.4.3 of Chap. 5.

Here, it is also important to note that any amendments made must be within the scope of subject matter that was disclosed in the initially filed patent application.

6.5.2.3 Amendments After Examination

Amendment of patent claims in light of objections received in the first examination report has been demonstrated by the empirical analysis to be critically important to applicants to assist them in excluding non-patentable subject matter at the examination stage. Patent applications that receive patentable subject matter-based objections are required to be amended or cancelled. If applicant amends objected patent claims, such amended claims would again be examined by the patent examiners against the subject matter provisions, and if the amended claims satisfy the subject matter requirements, they will be substituted in place of the objected original claims. However, if an examiner still believes that the amended version of claims includes non-patentable subject matter, the examiner may again raise patentable subject matter objections in the later reports until such claims are appropriately amended or cancelled. If the examination office received no response to the examination reports from the patent applicant in the prescribed period (noting that response times vary according to jurisdiction), such application would be considered abandoned by the relevant IP office. In view of this, it is advised that patent applicants may utilise the amendment option at the examination stage to amend the patent claims directed towards non-patentable subject matter, to overcome any persisting patentable subject matter-based objections.

6.6 Conclusion

Patent prosecution data related to patent applications with claims directed towards bioprinted tissues and bioinks has provided interesting insights to aid in understanding how the patent examiners of different jurisdictions are applying jurisdiction-specific patentable subject matter requirements and other patentability criteria on patent applications having the same subject matter. Combined with the results of the extended analysis on patentability criteria such as novelty and inventive step, the percentage of objections stemming from patent applications with claims directed towards bioprinted tissues or organs and methods is comparatively higher. This may be due to the fact that there is a large amount of prior art (including patents and non-patent literature) that has been extensively published in the last few years which was picked up by patent examiners during the novelty and inventive step analysis. This led to a greater number of novelty and inventive step based rejections for patent applications with claims directed towards bioprinted tissues or organs and methods, followed by patent applications with claims directed towards bioinks, hydrogels, biogels, and methods. The patentability criteria such as inventive step and disclosure requirements have become crucial in the patent examination process and in accepting or rejecting patent claims directed towards bioprinted tissues and bioinks as products and processes.

The conclusions of this chapter reveal that the patent prosecution data is a valuable tool to identify how patent examiners interpret and apply the existing patent law to emerging technologies such as bioprinting inventions, and particularly how different types of patent claims directed towards bioprinted tissues and bioinks are being assessed by patent examiners. Considering the overarching research question of this book, 'Is the patentable subject matter requirement a hurdle to patenting bioprinting inventions, specifically bioinks and bioprinted tissues,' in light of the conclusions of this chapter, there are two observations that may be made. The first is that the patentable subject matter requirement is a more significant hurdle in the EU than the other jurisdictions, followed by Australia and the US, in respect of patent applications with claims directed towards bioprinted tissues and bioinks. The second is that, as the patentable subject matter requirement can be a hurdle to patenting bioprinted tissues and bioinks, most patent applicants (with the help of experienced patent attorneys) appear deft at drafting patent claims to ensure they do not fall within the scope of patentable subject matter exceptions (such as products of nature and methods of treatment by surgery). This reduces the chances of incurring a patentable subject matter objection during examination. It can be concluded that the hurdle of the patentable subject matter requirement is apparently being overcome either through the drafting of patent claims prior to examination or through the amendment of objected claims after examination.

Chapter 7
Conclusion

The journey through the complexities of bioprinting and patent law brings us to a pivotal moment of reflection and anticipation. At the heart of this inquiry lies the fundamental question of how the legal frameworks designed to incentivise and protect innovation can adapt to technologies that straddle the line between the natural and the artificial. 3D bioprinting holds profound implications for healthcare, medicine, and biotechnology by biofabricating complex, functional human tissues and potentially organs. However, its intersection with patent law raises pressing questions about the extent to which existing legal principles can accommodate such radical innovation. This book's exploration of the patentable subject matter requirement and the unique challenges it poses for bioprinting has provided insights into the capacity and limitations of current patent regimes.

As outlined in the book's introductory chapters, patent law has historically evolved in response to technological advancements. From the earliest patents on mechanical inventions to the controversial patents on genetically modified organisms, each wave of innovation has expanded the boundaries of what can be considered patentable. The advent of bioprinting brings this evolution into sharp focus. The technology itself is rooted in principles of engineering, biology, and computer science and creates materials that are neither entirely natural nor entirely artificial, posing a challenge to traditional patent doctrines that exclude products of nature. Consequently, the questions explored in this book extend beyond the specifics of bioprinting to encompass broader issues about how the patent law interprets and defines invention in an era where the lines between human-made and natural processes are increasingly blurred.

One of the primary findings of this book is that patent law's role in the bioprinting field is multi-faceted. On one hand, patents provide essential protections that incentivise investment and research, especially in high-stakes fields like biotechnology, where the path from research to commercialisation is long and costly. By granting inventors a period of exclusive rights, patents encourage the development of novel bioprinting techniques, bioinks, and bioprinted tissues. Without this protection, companies and researchers may be reluctant to invest in bioprinting, potentially

P. Devarapalli, *Bioinked Boundaries*, https://doi.org/10.1007/978-981-96-2591-8_7

slowing the pace of innovation in a field with enormous potential to improve healthcare outcomes.

On the other hand, patents must also balance innovation with public access, a consideration that becomes especially salient in the context of healthcare-related technologies. As bioprinting progresses towards the creation of fully functional organs, the stakes surrounding patent protections will only grow. If too broad or restrictive, patents on bioprinting could limit access to life-saving technologies, particularly in cases where bioprinted tissues are used to develop personalised treatments or regenerative therapies. The risk, as discussed in this book, is that an overly expansive interpretation of patentable subject matter may lead to monopolisation in an area that holds the potential for wide-reaching public health benefits.

This book's comparative analysis of the United States, the Europe, and Australia highlights the diverse approaches taken by different jurisdictions in navigating this balance. Each of these regions brings its own legal provisions, ethical considerations, and policy priorities to bear on the question of bioprinting patents. The United States, with its more restrictive stance on patent eligibility following the *Mayo* and *Myriad* decisions, has taken a cautious approach to biotechnological patents that involve natural materials. In the US, if a patent claim directed towards a bioink includes additional elements that amount to significantly more than a judicial exception with 'markedly different characteristics,' such a claim may pass the Alice/Mayo test and be considered patentable. By contrast, the European Union's focus on ethical considerations and *ordre public*, as codified in the EPC, introduces additional layers of complexity, especially in relation to bioprinted tissues derived from stem cells. In Europe, patentable subject matter is determined primarily by reference to exceptions to patentability. A patent application with claims directed towards a bioprinted tissue or bioink that comprises embryonic stem cells (without presupposing a destructive use of human embryos) may be considered patentable in Europe. Australia's approach, which centres on the 'manner of manufacture' test, offers a relatively flexible framework that could allow for the patenting of bioprinting innovations, provided they demonstrate practical utility and significant human intervention. Particularly, the conclusions of the doctrinal analysis of Australian patent provisions indicated that subject matter related to bioinks and bioprinted tissues with claims directed towards a combination of natural and artificial elements such as synthetically produced 'living cells' or 'biomaterials' may meet the 'human action' and 'artificially created state of affairs' criteria in Australia.

Patent applications with claims directed towards a surgical method of bioprinting a tissue or bioink that can only be practised or performed by physicians or doctors may not be considered patentable in Australia and Europe but patentable in the US. However, a remedy for the infringement of such patented process is not available in the US (according to USC 287(c)(1) of AIA), effectively nullifying the right in the absence of enforcement. In providing these findings, this book also identified that the applicability of the patentable subject matter requirements of Australia, the US, and Europe vary on a case-to-case basis and highlighted how specific subject matter related to bioprinted tissues and bioinks may be claimed by patent applicants to overcome exceptions to patentability. While existing subject matter requirements (particularly the doctrinal limits established by Australian, US, and European patent laws and jurisprudence) may assist patent examiners in evaluating bioprinted tissues and bioinks, current provisions remain insufficient for addressing critical issues, such as 'identifying the level of human intervention' that is required for subject matter to be considered patentable. Through these comparative lenses, the analysis reveals that no single jurisdiction has fully resolved the issues surrounding bioprinting patents. Instead, each region grapples with the boundaries of patentable subject matter and the societal implications of biotechnological patents differently. While this diversity of approaches enriches the global patent landscape, it also creates a fragmented environment that poses challenges for inventors and companies seeking international protection for bioprinting technologies.

The empirical research presented in Chap. 6 sheds light on the practical challenges faced by patent applicants in the field of bioprinting. By analysing patent prosecution data, the study identified several trends that underscore the difficulties inherent in securing patent protection for bioprinted tissues and bioinks. One notable trend is the number of rejections based on subject matter eligibility, particularly in jurisdictions with strict interpretations of the product-of-nature exclusion. For instance, patent claims involving bioinks that closely resemble naturally occurring biological materials often face scrutiny under the current standards for patent eligibility, resulting in prolonged prosecution timelines and, in some cases, outright rejections.

The conclusions of Chap. 6 also indicated that most patent claims related to bioprinting technology are directed towards 'bioprinted tissue/organ and methods' followed by 'bioinks, hydrogels, biogels, and methods.' The empirical evidence also indicate that the patent offices are effectively using the currently available subject matter requirements derived from jurisdiction-specific patent laws and patent jurisprudence to assess patent claims related to bioprinted tissues and bioinks. However, while the currently available patent provisions of the respective jurisdictions are useful in assessing the patentability of bioprinted tissues and bioinks, there are still a number of outstanding issues. An example that may be considered is how a bioprinted tissue or organ having markedly different characteristics would replace a natural tissue or organ, which may raise issues in light of the enablement or sufficiency provisions. Given that enablement and sufficiency requirements are among the patentability requirements that must be met in order for a patent to be granted, it will be interesting to see how patent examiners handle such patent applications related to bioprinting tissues or organs that mimic their natural counterparts.

This empirical analysis also highlights the strategies that inventors and companies employ to navigate the complexities of bioprinting patents. By emphasising the synthetic components and engineered aspects of bioinks, applicants can strengthen their claims to demonstrate that their inventions involve significant human intervention. Additionally, some applicants are refining their claims to focus on specific applications of bioprinted tissues, such as their use in drug testing or disease modelling, which may increase the likelihood of satisfying the utility and inventive step requirements. These strategies underscore the adaptability of inventors within the constraints of existing patent law, even as they highlight the need for clearer guidelines tailored to the unique nature of bioprinting.

As the book has demonstrated, the existing patent framework is at a critical juncture and its evolution will play a crucial role in shaping the future of bioprinting. Patent law must continue to adapt if it is to remain relevant and effective in an era where technology increasingly challenges traditional legal definitions. In the case of bioprinting, this adaptation may involve rethinking the product-of-nature exclusion and developing specific guidelines for inventions that integrate biological and synthetic elements. Such an approach would enable patent systems to provide robust protections for bioprinting innovations while ensuring that these protections do not unduly restrict access to essential medical technologies.

Looking forward, the question of how patent law should evolve in response to bioprinting is not solely a legal or technical issue; it is also a question of values and priorities. Policymakers, courts, and patent offices must grapple with the ethical dimensions of bioprinting patents, particularly as the technology moves closer to the creation of human organs. In this context, the European Union's focus on ethical considerations within the EPC offers a valuable model for incorporating public interest concerns into the patentability analysis. However, even in jurisdictions without such explicit ethical clauses, there is scope for courts and patent offices to interpret patent laws in ways that reflect broader societal values, balancing the interests of inventors with the needs of the public.

The regulatory landscape for bioprinting will also benefit from international cooperation and harmonisation. As bioprinting technologies are inherently driven by cross-border global research collaborations and international markets, jurisdictions that work towards aligning their approaches to bioprinting patents will enhance legal certainty for innovators. Efforts towards harmonising patent standards, particularly regarding the patentability of bioinks and bioprinted tissues, could reduce the complexity and costs associated with securing patent protection across multiple regions. Such initiatives would also support the responsible dissemination of bioprinting technologies, facilitating access to life-saving innovations on a global scale.

The exploration of bioprinting and patent law presented in this book underscores the need for a responsive and equitable patent system that can accommodate the challenges and opportunities of emerging biotechnologies. The findings reveal a patent landscape that is both promising and precarious capable of fostering remarkable innovation yet constrained by doctrines that may not fully reflect the realities of twenty-first-century science. As bioprinting continues to progress, the pressure on patent law to evolve will only intensify. By addressing the legal and ethical dimensions of bioprinting patents, this book aims to contribute to an ongoing dialogue about the role of intellectual property in the age of biotechnology.

In closing, the potential of 3D bioprinting to transform medicine, improve patient outcomes, and address critical healthcare needs is undeniable. Yet realising this potential will require a patent system that is attuned to the unique attributes of bioprinting, one that reflects the rights of inventors with the public's need for access to life-enhancing technologies. By implementing thoughtful reforms, international cooperation, and a commitment to ethical considerations, patent law can continue to serve its foundational purpose of promoting scientific and technological progress for the benefit of society.

Bibliography

Articles/Conference Proceedings/Reports

Aboy, Mateo et al, 'Mayo's impact on patent applications related to biotechnology, diagnostics and personalized medicine' (2019) 37(5) *Nature Biotechnology* 513–518
Adesanya, Olumayowa O, 'Patenting bioprinted structures in a clime of moral uncertainty: time to amend the patents act?' (2019) 29(4) *Australian Intellectual Property Journal* 222–238
Ahrens, John et al, 'Programming Cellular Alignment in Engineered Cardiac Tissue via Bioprinting Anisotropic Organ Building Blocks' (2022) *Advanced Materials* 2200217
Althabhawi, Nabeel M and Zinatul Ashiqin Zainol, 'The Patent Eligibility of 3D Bioprinting: Towards a New Version of Living Inventions' Patentability' (2022) 12(1) *Biomolecules* 124
Ammar, Jamil, 'The Medical Mile Gearing toward 3D-Bespoke Healthcare: A Comparison of United States and European Union Patent Regimes' (2016) 52 *Gonzaga Law Reveiw* 279
An, J., C. K. Chua and V. Mironov, 'A perspective on 4D bioprinting' (2016) 2(1) *International Journal of Bioprinting* 3–5
Arslan-Yildiz, Ahu et al, 'Towards artificial tissue models: past, present, and future of 3D bioprinting' (2016) 8(1) *Biofabrication* 014103
Ashammakhi, N. et al, 'Advances and Future Perspectives in 4D Bioprinting' (2018) 13(12) *Biotechnology Journal*
Ashammakhi, N. et al, 'Bioinks and bioprinting technologies to make heterogeneous and biomimetic tissue constructs' (2019) 1 *Materials Today Bio*
Asif, E, 'Exclusion of diagnostic, therapeutic and surgical methods from patentability' (2013) 18 *Journal of Intellectual Property Rights*
Atala, Anthony, 'Tissue engineering in urology' (2001) 2(1) *Current urology reports* 83–92
Atala, A. and G. Forgacs, 'Three-Dimensional Bioprinting in Regenerative Medicine: Reality, Hype, and Future' (2019) 8(8) *Stem Cells Translational Medicine* 744–745
Auger, François A, Laure Gibot and Dan Lacroix, 'The pivotal role of vascularization in tissue engineering' (2013) 15(1) *Annual Review of Biomedical Engineering* 177–200
Barron, J. A. et al, 'Application of Laser Printing to Mammalian Cells' (2004) 453 *Thin Solid Films* 383–387
Barton, John H, 'Adapting the intellectual property system to new technologies' (1995) 10(2–3) *International Journal of Technology Management* 151–172
Bedell, Matthew L et al, 'A high-throughput approach to compare the biocompatibility of candidate bioink formulations' (2020) 17 *Bioprinting* e00068
Bertassoni, Luiz E, 'Bioprinting of complex multicellular organs with advanced functionality—recent progress and challenges ahead' (2022) 34(3) *Advanced Materials* 2101321

P. Devarapalli, *Bioinked Boundaries*, https://doi.org/10.1007/978-981-96-2591-8

Berthiaume, François, Timothy J Maguire and Martin L Yarmush, 'Tissue engineering and regenerative medicine: history, progress, and challenges' (2011) 2 *Annual review of chemical and biomolecular engineering* 403–430

Bhamare, Nilesh et al, '3D printing of human ear pinna using cartilage specific ink' (2021) 16(5) *Biomedical Materials* 055008

Bian, Liming et al, 'The influence of hyaluronic acid hydrogel crosslinking density and macromolecular diffusivity on human MSC chondrogenesis and hypertrophy' (2013) 34(2) *Biomaterials* 413–421

Binder, Kyle W et al, 'Drop-on-Demand Inkjet Bioprinting: A Primer' (2011) 6(01) *Gene Therapy and Regulation* 33–49

Bishop, E. S. et al, '3-D Bioprinting Technologies in Tissue Engineering and Regenerative Medicine: Current and Future Trends' (2017) 4(4) *Genes and Diseases* 185–195

Braga, Edimilson Junqueira et al, 'The role of specification in patent applications: A comparative study on sufficiency of disclosure' (2018) 53 *World Patent Information* 58–65

Campbell, Phil G and Lee E Weiss, 'Tissue Engineering with the Aid of Inkjet Printers' (2007) 7(8) *Expert Opinion on Biological Therapy* 1123–1127

Carley, Michael, Deepak Hedge and Alan Marco, 'What is the probability of receiving a US patent' (2015) 17 *Yale JL & Tech.* 203

Caulfield, Timothy et al, 'Evidence and anecdotes: an analysis of human gene patenting controversies' (2006) 24(9) *Nature Biotechnology* 1091–1094

Chakroun, Nefissa, 'Vagueness of patent claim language, claim construction and patent infringement–what a mess!' (2020) 69(11) *GRUR International* 1097–1104

Chambers, Jasemine, 'Patent Eligibility of Biotechnical Inventions in the United States, Europe, and Japan: How Much Patent Policy Is Public Policy' (2002) 34 *The George Washington International Law Review* 223

Chen, Jun-Long, 'Patent Eligibility Analysis of Bioprint Technology' (2017) 6(1) *NTUT Journal of Intellectual Property Law and Management* 46–66

Chiang, Tun-Jen, 'Defining patent scope by the novelty of the idea' (2011) 89 *Wash. UL Rev.* 1211

Chimene, D. et al, 'Advanced Bioinks for 3D Printing: A Materials Science Perspective' (2016) 44(6) *Annals of Biomedical Engineering* 2090–2102

Chopin-Doroteo, Mario, Edna Ayerim Mandujano-Tinoco and Edgar Krötzsch, 'Tailoring of the rheological properties of bioinks to improve bioprinting and bioassembly for tissue replacement' (2021) 1865(2) *Biochimica et Biophysica Acta (BBA)-General Subjects* 129782

Choudhury, D. et al, 'Organ-Derived Decellularized Extracellular Matrix: A Game Changer for Bioink Manufacturing?' (2018) 36(8) *Trends in Biotechnology* 787–805

Choudhury, Deepak, Shivesh Anand and May Win Naing, 'The arrival of commercial bioprinters–Towards 3D bioprinting revolution!' (2018) 4(2) *International Journal of Bioprinting*

Cockburn, Iain M, Samuel S Kortum and Scott Stern, 'Are all patent examiners equal? The impact of examiner characteristics' (Pt National Bureau of Economic Research Cambridge, Mass., USA) (2002)

Commission, Australian Law Reform, 'Genes and ingenuity: gene patenting and human health' (2004) *Commonwealth of Australia*

Crespi, R Stephen, 'The European Biotechnology Patent Directive is Dead' (1995) 13(5) *Trends in Biotechnology* 162–164

Cui, Xiaofeng et al, 'Thermal Inkjet Printing in Tissue Engineering and Regenerative Medicine' (2012) 6(2) *Recent Patents on Drug Delivery and Formulation* 149–155

Cui, Xiaofeng et al, 'Direct human cartilage repair using three-dimensional bioprinting technology' (2012) 18(11–12) *Tissue Engineering Part A* 1304–1312

Datta, P. et al, 'Essential Steps in Bioprinting: From Pre- to Post-bioprinting' (2018) 36(5) *Biotechnology Advances* 1481–1504

De Jong, Jos et al, 'Air Entrapment in Piezo-Driven Inkjet Printheads' (2006) 120(3) *The Journal of the Acoustical Society of America* 1257–1265

De Moor, Lise et al, 'Engineering microvasculature by 3D bioprinting of prevascularized spheroids in photo-crosslinkable gelatin' (2021) 13(4) *Biofabrication* 045021

De Santis, Martina M et al, 'Extracellular-matrix-reinforced bioinks for 3D bioprinting human tissue' (2021) 33(3) *Advanced materials* 2005476

Dent, Chris, "Generally inconvenient': the 1624 Statute of Monopolies as political compromise' (2009) 33(2) *Melbourne University Law Review* 415–453

Dey, Madhuri and Ibrahim T Ozbolat, '3D bioprinting of cells, tissues and organs' (2020)

Dhariwala, Busaina, Elaine Hunt and Thomas Boland, 'Rapid prototyping of tissue-engineering constructs, using photopolymerizable hydrogels and stereolithography' (2004) 10(9–10) *Tissue engineering* 1316–1322

Diwan, Ishac and Dani Rodrik, 'Patents, appropriate technology, and North-South trade' (1991) 30(1–2) *Journal of International economics* 27–47

Dolmage, Brieanna, 'The Evolution of Patentable Subject Matter in the United States' (2005) 27 *Whittier Law Review* 1023

Duan, Bin et al, 'Three-dimensional printed trileaflet valve conduits using biological hydrogels and human valve interstitial cells' (2014) 10(5) *Acta biomaterialia* 1836–1846

Duffy, John F, 'Rules and Standards on the Forefront of Patentability' (2009) 51 *William & Mary Law Review* 609

Ebrahim, Tabrez Y, '3D Bioprinting Patentable Subject Matter Boundaries' (2017) 41 *Seattle University Law Review* 1

Eisenberg, Rebecca S, 'Why the gene patenting controversy persists' (2002) 77(12) *Academic Medicine* 1381–1387

Encaoua, David, Dominique Guellec and Catalina Martínez, 'Patent systems for encouraging innovation: Lessons from economic analysis' (2006) 35(9) *Research policy* 1423–1440

Eswaramoorthy, S. D., S. Ramakrishna and S. N. Rath, 'Recent Advances in Three-Dimensional Bioprinting of Stem Cells' (2019) 13(6) *Journal of Tissue Engineering and Regenerative Medicine* 908–924

Eswaramoorthy, Sindhuja D et al, '3D bioprinting of mesenchymal stem cells and endothelial cells in an alginate-gelatin-based bioink' (2021) 5(1) *Journal of 3D printing in medicine* 23–36

Filardo, Giuseppe et al, 'Patient-specific meniscus prototype based on 3D bioprinting of human cell-laden scaffold' (2019) 8(2) *Bone & joint research* 101–106

Frakes, Michael D and Melissa F Wasserman, 'Does the US patent and trademark office grant too many bad patents: evidence from a quasi-experiment' (2015) 67 *Stan. L. Rev.* 613

Fromer, Jeanne C, 'The Layers of Obviousness in Patent Law' (2008) 22 *Harvard Journal of Law & Technology* 75

Gao, G. F. et al, 'Improved Properties of Bone and Cartilage Tissue from 3D Inkjet-Bioprinted Human Mesenchymal Stem Cells by Simultaneous Deposition and Photocrosslinking in PEG-GelMA' (2015) 37(11) *Biotechnology Letters* 2349–2355

Gao, B. et al, '4D Bioprinting for Biomedical Applications' (2016) 34(9) *Trends in Biotechnology* 746–756

Gawas, Vijay M, 'Doctrinal legal research method a guiding principle in reforming the law and legal system towards the research development' (2017)

Gold, E Richard and Alain Gallochat, 'The European biotech directive: past as prologue' (2001) 7(3) *European Law Journal* 331–366

Gopinathan, J. and I. Noh, 'Recent Trends in Bioinks for 3D Printing' (2018) 22 *Biomaterials Research* 11–11

Govindharaj, Mano et al, '3D Bioprinting of human Mesenchymal Stem Cells in a novel tunic decellularized ECM bioink for Cartilage Tissue Engineering' (2022) *Materialia* 101457

Greenfield, Debra L, 'Greenberg v. Miami Children's Hospital: Unjust enrichment and the patenting of human genetic material' (2006) 15 *Annals Health L.* 213

Grigoryan, B. et al, 'Multivascular Networks and Functional Intravascular Topologies within Biocompatible Hydrogels' (2019) 364(6439) *Science* 458–+

Gu, Qi et al, 'Stem Cell Bioprinting: Functional 3D Neural Mini-Tissues from Printed Gel-Based Bioink and Human Neural Stem Cells' (2016) 5(12) *Advanced healthcare materials* 1428–1428

Gu, Qi et al, '3D bioprinting human induced pluripotent stem cell constructs for in situ cell proliferation and successive multilineage differentiation' (2017) 6(17) *Advanced healthcare materials* 1700175

Guillemot, F. et al, 'High-Throughput Laser Printing of Cells and Biomaterials for Tissue Engineering' (2010) 6(7) *Acta Biomaterialia* 2494–2500

Guillemot, Fabien et al, 'Laser-assisted bioprinting to deal with tissue complexity in regenerative medicine' (2011) 36(12) *Mrs Bulletin* 1015–1019

Guillotin, B. et al, 'Laser Assisted Bioprinting of Engineered Tissue with High Cell Density and Microscale Organization' (2010) 31(28) *Biomaterials* 7250–7256

Gungor-Ozkerim, P Selcan et al, 'Bioinks for 3D bioprinting: an overview' (2018) 6(5) *Biomaterials science* 915–946

Han, Yu Long et al, 'BioPen: direct writing of functional materials at the point of care' (2014) 4(1) *Scientific reports* 1–5

Holman, Molly A and Stephen R Munzer, 'Intellectual property rights in genes and gene fragments: A registration solution for expressed sequence tags' (1999) 85 *Iowa Law Review* 735

Holzl, K. et al, 'Bioink Properties Before, During and After 3D bioprinting' (2016) 8(3) *Biofabrication*

Hopp, B. et al, 'Survival and Proliferative Ability of Various Living Cell Types after Laser-Induced Forward Transfer' (2005) 11(11–12) *Tissue Engineering* 1817–1823

Horejs, Christine, 'Bioprinted mini kidneys' (2021) 6(1) *Nature Reviews Materials* 6–6

Hospodiuk, M. et al, 'The Bioink: A Comprehensive Review on Bioprintable Materials' (2017) 35(2) *Biotechnology Advances* 217–239

Hoyt, A Sasha, 'The Impact of Uncertainty Regarding Patent Eligible Subject Matter for Investment in US Medical Diagnostic Technologies' (2022) 79(1) *Washington and Lee Law Review* 397

Hull, Sarah M, Lucia G Brunel and Sarah C Heilshorn, '3D bioprinting of cell-laden hydrogels for improved biological functionality' (2022) 34(2) *Advanced Materials* 2103691

Hutchinson, Terry, 'The doctrinal method: Incorporating interdisciplinary methods in reforming the law' (2015) 8 *Erasmus L. Rev.* 130

Inventive step and raising the bar further?: A proposal to assist examiners, 107) (Intellectual Property Forum: journal of the Intellectual and Industrial Property Society of Australia and New Zealand) annex 0815–2098

Isaacson, Abigail, Stephen Swioklo and Che J Connon, '3D bioprinting of a corneal stroma equivalent' (2018) 173 *Experimental eye research* 188–193

Jamroz, W. et al, '3D Printing in Pharmaceutical and Medical Applications - Recent Achievements and Challenges' (2018) 35(9) *Pharmaceutical Research*

Jayasinghe, Suwan N, Amer N Qureshi and Peter AM Eagles, 'Electrohydrodynamic jet processing: an advanced electric-field-driven jetting phenomenon for processing living cells' (2006) 2(2) *Small* 216–219

Jeon, Oju et al, 'Stem cell-laden hydrogel bioink for generation of high resolution and fidelity engineered tissues with complex geometries' (2022) 15 *Bioactive materials* 185–193

Ji, Shen and Murat Guvendiren, 'Recent Advances in Bioink Design for 3D Bioprinting of Tissues and Organs' (2017) 5 *Frontiers in Bioengineering and Biotechnology* 23–23

Ji, S., E. Almeida and M. Guvendiren, '3D Bioprinting of Complex Channels within Cell-Laden Hydrogels' (2019) 95 *Acta Biomaterialia* 214–224

Jiang, Tao et al, 'Extrusion bioprinting of soft materials: An emerging technique for biological model fabrication' (2019) 6(1) *Applied Physics Reviews* 011310

Jovic, Thomas H et al, '3D Bioprinting and the Future of Surgery' (2020) *Frontiers in Surgery* 129

Jürgens, Björn and Victor Herrero-Solana, 'Espacenet, Patentscope and Depatisnet: A comparison approach' (2015) 42 *World Patent Information* 4–12

Kamariah, Ismail, Zaidi Wan Omar Wan and Abdul Majid Izaidin, 'The commercialisation process of patents by universities' (2011) 5(17) *African Journal of Business Management* 7198–7208

Kaufman, Matthew H., 'Parthenogenetic Activation of Oocytes' (2018) 2018(1) *Cold Spring Harbor Protocols*

Keriquel, Virginie et al, 'In situ Printing of Mesenchymal Stromal Cells, by Laser-Assisted Bioprinting, for In vivo Bone Regeneration Applications' (2017) 7 *Scientific Reports*

Kica, Evisa and Nico Groenendijk, 'The European patent system: dealing with emerging technologies' (2011) 24(1–2) *Innovation: The European journal of social science research* 85–105

Kim, Byoung Soo et al, 'Decellularized Extracellular Matrix: A Step Towards the Next Generation Source for Bioink Manufacturing' (2017) 9(3) *Biofabrication*

Kirillova, A. et al, '4D Biofabrication Using Shape-Morphing Hydrogels' (2017) 29(46) *Advanced Materials*

Kizawa, Hideki et al, 'Scaffold-free 3D bio-printed human liver tissue stably maintains metabolic functions useful for drug discovery' (2017) 10 *Biochemistry and biophysics reports* 186–191

Klebe, Robert J, 'Cytoscribing: a method for micropositioning cells and the construction of two-and three-dimensional synthetic tissues' (1988) 179(2) *Experimental cell research* 362–373

Kolesky, D. B. et al, '3D Bioprinting of Vascularized, Heterogeneous Cell-Laden Tissue Constructs' (2014) 26(19) *Advanced Materials* 3124–3130

Kolesky, David B et al, 'In vitro human tissues via multi-material 3-D bioprinting' (2018) 46(4) *Alternatives to Laboratory Animals* 209–215

Kumar, S. A. et al, 'A Visible Light Crosslinkable, Fibrin-Gelatin Based Bioprinted Cardiac Construct with Human Cardiomyocytes and Fibroblasts' (2019) *ACS Biomaterials Science and Engineering*

Kuo, C. Y. et al, 'Repair of Tympanic Membrane Perforations with Customized Bioprinted Ear Grafts Using Chinchilla Models' (2018) 24(5–6) *Tissue Engineering Part A* 527–535

Lacy, Patricia A, 'Gene patenting: universal heritage vs. reward for human effort' (1998) 77 *Or. L. Rev.* 783

Lai, J. C., 'D'Arcy v. Myriad Genetics: A Demand for the "Made"'or "Non-Information"'and Clear Subject Matter?' (2016) 47(5) *IIC-International Review of Intellectual Property and Competition Law* 537–568

Lam, T. et al, 'Photopolymerizable Gelatin and Hyaluronic Acid for Stereolithographic 3D Bioprinting of Tissue-Engineered Cartilage' (2019) (2019/03/13) *The Journal of Biomedical Materials Research*

Lawlor, Kynan T et al, 'Cellular extrusion bioprinting improves kidney organoid reproducibility and conformation' (2021) 20(2) *Nature materials* 260–271

Ld, Co Ld v Cocker Brothers, 'Less than obvious: the possible implications of the Productivity Commission's proposed changes to inventive step in Australia' (2016)

Lemley, Mark A and Bhaven Sampat, 'Examiner characteristics and patent office outcomes' (2012) 94(3) *Review of economics and statistics* 817–827

Levidow, Les et al, 'Regulating agricultural biotechnology in Europe: harmonisation difficulties, opportunities, dilemmas' (1996) 23(3) *Science and Public Policy* 135–157

Levin, R., 'A Patent System for the 21st Century' (2004) 20(4) *Issues in Science and Technology* 49–54

Li, Jipeng et al, 'Recent Advances in Bioprinting Techniques: Approaches, Applications and Future Prospects' (2016) 14(1) *Journal of translational medicine* 271

Li, Y. C. et al, '4D Bioprinting: The Next-Generation Technology for Biofabrication Enabled by Stimuli-Responsive Materials' (2017) 9(1) *Biofabrication*

Li, Hongbin et al, 'Handheld bioprinting strategies for in situ wound dressing' (2021) 65(3) *Essays in biochemistry* 533–543

Lichtman, Douglas, 'Rethinking prosecution history estoppel' (2004) *The University of Chicago Law Review* 151–182

Lim, Pheh Hoon and Phoebe Li, 'Patentability of biofabricated human organs: products of nature or products derived from nature revisited', *Patenting Biotechnical Innovation* (Edward Elgar Publishing, 2022) 69–88

Little, Christopher James, Nahshon Kenneth Bawolin and Xiongbiao Chen, 'Mechanical properties of natural cartilage and tissue-engineered constructs' (2011) 17(4) *Tissue Engineering Part B: Reviews* 213–227

Liu, Justin et al, 'Rapid 3D bioprinting of in vitro cardiac tissue models using human embryonic stem cell-derived cardiomyocytes' (2019) 13 *Bioprinting* e00040

Lu, Qiang, Amanda Myers and Scott Beliveau, 'USPTO patent prosecution research data: Unlocking office action traits' (2017) *Economic Working Paper Series*

Mandrycky, Christian et al, '3D bioprinting for engineering complex tissues' (2016) 34(4) *Biotechnology advances* 422–434

Mao, Jinshu et al, 'Study of novel chitosan-gelatin artificial skin in vitro' (2003) 64(2) *Journal of Biomedical Materials Research Part A: An Official Journal of The Society for Biomaterials, The Japanese Society for Biomaterials, and The Australian Society for Biomaterials and the Korean Society for Biomaterials* 301–308

Marco, Alan C, Joshua D Sarnoff and AW Charles, 'Patent claims and patent scope' (2019) 48(9) *Research Policy* 103790

Martello, Graziano and Austin Smith, 'The nature of embryonic stem cells' (2014) 30 *Annual review of cell and developmental biology* 647–675

Matai, Ishita et al, 'Progress in 3D Bioprinting Technology for Tissue/Organ Regenerative Engineering' (2020) 226 *Biomaterials* 119536–119536

Mazzoleni, Roberto and Richard R Nelson, 'The benefits and costs of strong patent protection: a contribution to the current debate' (1998) 27(3) *Research policy* 273–284

Melchels, F. P. W., J. Feijen and D. W. Grijpma, 'A Review on Stereolithography and its Applications in Biomedical Engineering' (2010) 31(24) *Biomaterials* 6121–6130

Miao, S. D. et al, 'Stereolithographic 4D Bioprinting of Multiresponsive Architectures for Neura Engineering' (2018) 2(9) *Advanced Biosystems*

Minssen, Timo and Marc Mimler, 'Patenting Bioprinting-Technologies in the US and Europe–The 5th Element in the 3rd Dimension' (2016)

Mirabel, Eric P, 'Product-by-process claims: a practical perspective' (1986) 68 *Journal of the Patent and Trademark Office Society* 3

Mirani, Bahram et al, 'Microfluidic 3D printing of a photo-cross-linkable bioink using insights from computational modeling' (2021) 7(7) *ACS Biomaterials Science & Engineering* 3269–3280

Mirdamadi, Eman et al, 'FRESH 3D bioprinting a full-size model of the human heart' (2020) 6(11) *ACS Biomaterials Science & Engineering* 6453–6459

Miri, A. K. et al, 'Microfluidics-Enabled Multimaterial Maskless Stereolithographic Bioprinting' (2018) 30(27) *Advanced Materials*

Moir, Hazel VJ, 'Empirical evidence on the inventive step' (2013) *European Intellectual Property Review*

Murphy, S. V. and A. Atala, '3D Bioprinting of Tissues and Organs' (2014) 32(8) (2014/08/06) *Nature Biotechnology* 773–85

Murphy, Sean V and Anthony Atala, '3D bioprinting of tissues and organs' (2014) 32(8) *Nature biotechnology* 773–785

Naghieh, Saman and Xiongbiao Chen, 'Printability–A key issue in extrusion-based bioprinting' (2021) 11(5) *Journal of Pharmaceutical Analysis* 564–579

Nakamura, Makoto et al, 'Biocompatible Inkjet Printing Technique for Designed Seeding of Individual Living Cells' (2005) 11(11–12) *Tissue Engineering* 1658–1666

Ng, Hooi Yee et al, 'Bioprinting of artificial blood vessels' (2018) 4(2) *International Journal of Bioprinting*

Ng, Wei Long, Chee Kai Chua and Yu-Fang Shen, 'Print Me An Organ! Why We Are Not There Yet' (2019) 97 *Progress in Polymer Science*

Nicolai, Thomas R, 'The European Patent Convention: A Theoretical and Practical Look at International Legislation' (Pt HeinOnline) (1971) 5 *International Lawyer* 135

Norotte, Cyrille et al, 'Scaffold-free vascular tissue engineering using bioprinting' (2009) 30(30) *Biomaterials* 5910–5917

Noruzi, Alireza and Mohammadhiwa Abdekhoda, 'Google Patents: The global patent search engine' (2014) 11(1) *Webology*

Odde, David J and Michael J Renn, 'Laser-guided direct writing for applications in biotechnology' (1999) 17(10) *Trends in biotechnology* 385–389

Odde, D. J. and M. J. Renn, 'Laser-Guided Direct Writing of Living Cells' (2000) 67(3) *Biotechnology and Bioengineering* 312–318

Opinion of The European Group on Ethics in Science and New Technologies to the European Commission - Ethical Aspects of Patenting Inventions Involving Human Stem Cells (The European Group on Ethics in Science and New Technologies, 07 May 2002)

Organ printing: a novel tissue engineering paradigm) (5th European Conference of the International Federation for Medical and Biological Engineering)

Osidak, Egor O. et al, 'Viscoll Collagen Solution as a Novel Bioink for Direct 3D Bioprinting' (2019) 30(3) *Journal of Materials Science-Materials in Medicine*

Ouyang, Liliang et al, 'Three-dimensional bioprinting of embryonic stem cells directs highly uniform embryoid body formation' (2015) 7(4) *Biofabrication* 044101

Ozbolat, Ibrahim T, 'Scaffold-Based or Scaffold-Free Bioprinting: Competing or Complementing Approaches?' (2015) 6(2) *Journal of Nanotechnology in Engineering and Medicine* 024701

Parikh, Meghana Tushar, 'Unleashing bioprinting technology through patent intelligence' (2021) 26(6) *Drug Discovery Today* 1547–1555

Park, H. K. et al, 'A Visible Light-Curable yet Visible Wavelength-Transparent Resin for Stereolithography 3D Printing' (2018) 10 *Nature Publishing Group Asia Materials*

Patent litigation in the United States- an overview) (Sohn International Symposium; Advanced Processing of Metals and Materials Volume 9: Legal, Management and Environmental Issues)

Pazhouhnia, Zahra et al, 'Portable hand-held bioprinters promote in situ tissue regeneration' (2022) *Bioengineering & Translational Medicine* e10307

Peng, Weijie et al, '3D Bioprinting for Drug Discovery and Development in Pharmaceutics' (2017) 57 *Acta biomaterialia* 26–46

Peng, Weijie et al, 'Challenges in bio-fabrication of organoid cultures' (2018) *Cell Biology and Translational Medicine, Volume 3* 53–71

Pepper, Matthew E et al, 'Cell Settling Effects on A Thermal Inkjet Bioprinter' 2011 Annual International Conference of the IEEE Engineering in Medicine and Biology Society)

Pepper, M. E. et al, 'Characterizing the Effects of Cell Settling on Bioprinter Output' (2012) 4(1) *Biofabrication*

Pera, Martin F, Benjamin Reubinoff and Alan Trounson, 'Human embryonic stem cells' (2000) 113(1) *Journal of cell science* 5–10

Petherbridge, Lee and R Polk Wagner, 'The Federal Circuit and Patentability: An Empirical Assessment of the Law of Obviousness' (2006) 85 *Texas Law Review* 2051

Pila, Justine, 'Inherent patentability in Anglo-Australian law: A history' (2003) 14 *Australian Intellectual Property Journal* 109–166

Plomer, Aurora et al, 'Stem cell patents: European patent law and ethics report' (2006) *FP6 Life Sciences, Genomics and Biotechnology for Health', SSA LSSB-CT-2004–005251. Nottingham: School of Law*

Plomer, Aurora, Kenneth S Taymor and Christopher Thomas Scott, 'Challenges to human embryonic stem cell patents' (2008) 2(1) *Cell Stem Cell* 13–17

Ramaiahgari, S. C. et al, 'A 3D In vitro Model of Differentiated HepG2 Cell Spheroids with Improved Liver-like Properties for Repeated Dose High-Throughput Toxicity Studies' (2014) 88(5) *Archives of Toxicology* 1083–1095

Report from the Commission to the European Parliament and the Council - Development and implications of patent law in the field of biotechnology and genetic engineering (Commission of The European Communities No 52002DC0545, 07/10/2002)

Report from the Commission to the the Council and European Parliament - Development and implications of patent law in the field of biotechnology and genetic engineering (Commission of The European Communities No 52005DC0312, 14/07/2005)

Review of intellectual property legislation under the Competition Principles Agreement–Final Report Canberra: Commonwealth of Australia, *IP Australia*,

Rich, Giles S, 'Principles of Patentability' (1959) 28 *The George Washington Law Review* 393

Rogers, Arthur, 'European Union approves gene-patent Directive' (1998) 351(9114) *The Lancet* 1500

Satpathy, Aishwarya et al, 'Developments with 3D bioprinting for novel drug discovery' (2018) 13(12) *Expert opinion on drug discovery* 1115–1129

Sauer, Igor M, Nathanael Raschzok and Peter Neuhaus, 'The artificial liver, in vivo tissue engineering, and organ printing' (2014) *Textbook of organ transplantation* 545–553

Saunders, Rachel Elizabeth and Brian Derby, 'Inkjet printing biomaterials for tissue engineering: bioprinting' (2014) 59(8) *International Materials Reviews* 430–448

Schiele, Nathan R. et al, 'Laser-Based Direct-Write Techniques for Cell Printing' (2010) 2(3) *Biofabrication*

Sherkow, Jacob S and Henry T Greely, 'The history of patenting genetic material' (2015) 49 *Annual Review of Genetics* 161–182

Sherman, Brad, 'Before the High Court D'Arcy v Myriad Genetics Inc: Patenting Genes in Australia' (2015) 37(1) *Sydney Law Review* 135–146

Shigemasa, Yoshihiro and Saburo Minami, 'Applications of chitin and chitosan for biomaterials' (1996) 13(1) *Biotechnology and Genetic Engineering Reviews* 383–420

Skardal, A., J. Zhang and G. D. Prestwich, 'Bioprinting Vessel-Like Constructs using Hyaluronan Hydrogels Crosslinked with Tetrahedral Polyethylene Glycol Tetracrylates' (2010) 31(24) *Biomaterials* 6173–6181

Skirvin, Robert M, 'Natural and induced variation in tissue culture' (1978) 27(1) *Euphytica* 241–266

Solis, L. H. et al, 'Thermal Inkjet Bioprinting Triggers The Activation Of The VEGF Pathway in Human Microvascular Endothelial Cells In vitro' (2019) 11(4) (2019/06/01) *Biofabrication* 045005

Sorkio, Anni et al, 'Human stem cell based corneal tissue mimicking structures using laser-assisted 3D bioprinting and functional bioinks' (2018) 171 *Biomaterials* 57–71

Spencer, Richard, 'The German Patent Office' (1949) 31 *Journal of the Patent and Trademark Office Society* 79

Spillmann-Fürst, Ingeborg, 'Europe edges closer to GMO patent harmonization' (1999) 17(9) *Nature Biotechnology* 842–843

Sterne, Robert Greene and Lawrence B Bugaisky, 'The Expansion of Statutory Subject Matter Under the 1952 Patent Act' (2004) 37 *Akron Law Review* 217

Stolbova, Ekaterina and Federica Baldan, 'Abstract ideas and patent-eligible subject matter in the US patent system: who can provide guidance in this uncertain legal landscape?' (2021) 16(11) *Journal Of Intellectual Property Law and Practice* 1229–1241

Strobel, Hannah A, Sarah M Moss and James B Hoying, 'Mechanical considerations of bioprinted tissue' (2020) 6 *Frontiers in Mechanical Engineering* 568726

Thums, Doris, 'Patent Protection for Medical Treatment--A Distinction Between Patent and Medical Law' (1996) 27(4) *IIC: International Review of Industrial Property and Copyright Law* 423–447

Tibbits, S., '4D Printing: Multi-Material Shape Change' (2014) 84(1) *Architectural Design* 116–121

Tiwari, Arpita P et al, 'Bioink: a 3D-bioprinting tool for anticancer drug discovery and cancer management' (2021) 26(7) *Drug Discovery Today* 1574–1590

Tomasina, Clarissa et al, 'Bioprinting vasculature: materials, cells and emergent techniques' (2019) 12(17) *Materials* 2701

Tvedt, Morten Walløe and Ellen-Marie Forsberg, 'The room for ethical considerations in patent law applied to biotechnology' (2017) 20(5–6) *The Journal of World Intellectual Property* 160–177

Van Caenegem, William, 'The Technicality Requirement, Patent Scope and Patentable Subject Matter in Australia' (2002) 13(1) *Australian Intellectual Property Journal* 309–327

Van Overwalle, Geertrui, 'The implementation of the biotechnology directive in Belgium and its aftereffects: the introduction of a new research exemption and a compulsory license for public health' (2006) 37 *International Review of Intellectual Property and Competition Law (IIC)* 889–920

Vandael, Eline et al, 'Risk factors for QTc-prolongation: systematic review of the evidence' (2017) 39(1) *International journal of clinical pharmacy* 16–25

Vaver, David, 'The Perplexities of Patent Prosecution History: Procedure over Principle?' (2021) *Osgoode Legal Studies Research Paper Forthcoming*

Ventose, Eddy D, 'Disclaiming methods of medical treatment under the European Patent Convention' (2009) 4(10) *Journal of Intellectual Property Law & Practice* 706–713

Verdonk, Peter CM et al, 'Characterisation of human knee meniscus cell phenotype' (2005) 13(7) *Osteoarthritis and Cartilage* 548–560

Walterscheid, Edward C, 'Novelty in Historical Perspective (Part I)' (1993) 75 *Journal of the Patent and Trademark Office Society* 689

Wang, Xiaohong, Boaz Lloyd Rijff and Gilson Khang, 'A building-block approach to 3D printing a multichannel, organ-regenerative scaffold' (2017) 11(5) *Journal of Tissue Engineering and Regenerative Medicine* 1403–1411

Wang, Z. et al, '3D bioprinted functional and contractile cardiac tissue constructs' (2018) 70 *Acta Biomaterialia* 48–56

Wang, Yuzhen et al, 'Tailoring bioinks of extrusion-based bioprinting for cutaneous wound healing' (2022) *Bioactive Materials*

Whitworth, Laura C, 'Comparison of the Implementation of Statutory Patent Eligibility Requirements Applied to Gene Patents in the European Union, the United States, and Australia' (2016) 56 *IDEA* 449

Xu, T. et al, 'Inkjet Printing of Viable Mammalian Cells' (2005) 26(1) *Biomaterials* 93–99

Xu, Feng et al, 'Embryonic stem cell bioprinting for uniform and controlled size embryoid body formation' (2011) 5(2) *Biomicrofluidics* 022207

Xu, Jie et al, 'Advances in the research of bioinks based on natural collagen, polysaccharide and their derivatives for skin 3D bioprinting' (2020) 12(6) *Polymers* 1237

Yerneni, Saigopalakrishna S et al, 'Bioprinting Exosome-like Extracellular Vesicle Microenvironments' (2019) 13 *Bioprinting* e00041

Ying, G. L. et al, 'Aqueous Two-Phase Emulsion Bioink-Enabled 3D Bioprinting of Porous Hydrogels' (2018) 30(50) *Advanced Materials*

Yoo, Seung-Schik, '3D-Printed Biological Organs: Medical Potential and Patenting Opportunity' (2015) 25(5) *Expert Opinion on Therapeutic Patents* 507–11

Ze, Yiting et al, 'Biodegradable Inks in Indirect Three-Dimensional Bioprinting for Tissue Vascularization' (2022) *Frontiers in Bioengineering and Biotechnology* 461

Zhang, A. P. et al, 'Rapid Fabrication of Complex 3D Extracellular Microenvironments by Dynamic Optical Projection Stereolithography' (2012) 24(31) *Advanced Materials* 4266–+

Zhang, Yu Shrike et al, '3D extrusion bioprinting' (2021) 1(1) *Nature Reviews Methods Primers* 1–20

Zheng, Zizhuo et al, 'Visible light-induced 3D bioprinting technologies and corresponding bioink materials for tissue engineering: a review' (2021) 7(7) *Engineering* 966–978

Zhou, D. et al, 'Bioinks for Jet-based Bioprinting' (2019) *Bioprinting* e00060

Zhuang, Ziyun and Ho Pui Ho, 'Application of digital micromirror devices (DMD) in biomedical instruments' (2020) 13(06) *Journal of Innovative Optical Health Sciences* 2030011

Zohora, Fatema Tuj and Abu Yousuf Mohammad Anwarul Azim, 'Inkjet Printing: An Emerging Technology for 3D Tissue or Organ Printing' (2014) 10(30) *European Scientific Journal*

Books/Book Chapters

Bártolo, Paulo Jorge, 'Stereolithographic Processes', *Stereolithography* (Springer, 2011) 1–36

Camacho, P. et al, 'Materials as Bioinks and Bioink Design', *3D Bioprinting in Medicine* (Springer, 2019) 67–100

Council, National Research, *Patents in the knowledge-based economy* (National Academies Press, 2003)

Devillard, Raphael et al, 'Cell Patterning by Laser-Assisted Bioprinting' in M. Piel and M. Thery (eds), *Micropatterning in Cell Biology, Pt A* (2014) vol 119, 159–174

Freeman, Christopher et al, *As time goes by: from the industrial revolutions to the information revolution* (Oxford University Press, 2001)

Hussain, Y. and J. T. Butcher, 'Bioprinting Cardiovascular Organs', *3D Printing Applications in Cardiovascular Medicine* (Elsevier, 2018) 163–187

Kérourédan, Olivia et al, 'Laser-Assisted Bioprinting for Tissue Engineering', *Biomaterials and Nanotechnology for Tissue Engineering* (CRC Press, 2016) 287–304

Kumar, Hitendra and Keekyoung Kim, 'Stereolithography 3D bioprinting', *3D Bioprinting* (Springer, 2020) 93–108

Li, Phoebe H., '3D bioprinting: Regulation, Innovation, and Patents', *3D Bioprinting for Reconstructive Surgery* (2018) 217–231

Mireles, Michael S, 'Patent-Eligible Subject Matter in the United States: An Evolving Landscape', *Intellectual Property Issues in Microbiology* (Springer, 2019) 3–35

Pepper, M. E. et al, 'Cell Settling Effects on a Thermal Inkjet Bioprinter', *2011 Annual International Conference of the Ieee Engineering in Medicine and Biology Society* (2011) 3609–3612

Prochazkova, Michaela et al, 'Embryonic versus adult stem cells', *Stem cell biology and tissue engineering in dental sciences* (Elsevier, 2015) 249–262

Raman, Ritu and Rashid Bashir, 'Stereolithographic 3D Bioprinting for Biomedical Applications', *Essentials of 3D Biofabrication and Translation* (Elsevier, 2015) 89–121

Rosser, Jordan and Daniel J Thomas, 'Bioreactor processes for maturation of 3D bioprinted tissue', *3D Bioprinting for reconstructive surgery* (Elsevier, 2018) 191–215

Seville, Catherine, *EU intellectual property law and policy* (Edward Elgar Publishing, 2009)

Straus, Joseph, 'Ordre public and morality issues in patent eligibility', *Intellectual property in common law and civil law* (Edward Elgar Publishing, 2013)

Twain, Mark, *A Connecticut Yankee in King Arthur's Court* (New York,, London,: Harper & brothers, 1917)

Cases

Alice Corporation Pty Ltd v CLS Bank International, 134 US 2347 (2014)

American Fruit Growers, Inc. v. Brogdex Co., 283 U. S. 1, 11 [1931]

American Wood-Paper Co v Fibre Disintegrating Co, [1874] USSC 189

Anaesthetic supplies pty limited v. Rescare limited [1994] 50 FCR 1

Apotex Pty Ltd v Sanofi-Aventis Australia Pty Ltd [2013] HCA 50

Ariosa Diagnostics Inc v Sequenom Inc [2021] FCAFC 101

Arrowhead Research Corporation [2016] APO 70

Association for Molecular Pathology v Myriad Genetics Inc, 133 US 2107 [2013]

Association for Molecular Pathology v. Myriad Genetics, Inc., 132 S.Ct. 1794 [2012]

Axiogenesis AG v GE Healthcare UK Limited [2019] T0385/14

Bergy, 571 F.2d 40 (C.C.P.A. 1978)

Bilski v Kappos, 561 US 593 (2010)

Breast and Ovarian Cancer/University of Utah T-1213/05 [2007]

Bristol-Myers Squibb Co v F H Faulding & Co Ltd [2000] FCA 316
Cancer Voices Australia v Myriad Genetics Inc [2013] FCA 65
Cargill Incorporated v Dow AgroSciences LLC [2016] APO 43
CCOM Pty Ltd v Jiejing Pty Ltd (1994) 51 FCR 260
Christian Franceries/Traffic Regulation ('Christian Franceries') T16/83 [1988] EPOR
Commissioner of Patents v RPL Central Pty Ltd (2015) 238 FCR 27; [2015] FCAFC 177
Culturing Stem Cells/Technion T 2221/10 [2014]
D'Arcy v Myriad Genetics Inc [2014] FCAFC 115
D'Arcy v Myriad Genetics Inc [2015] HCA 35
Diamond v. Chakrabarty, 447 U.S. 303, 309 [1980]
Diamond v. Diehr, 450 U. S. 175 [1981]
Funk Bros. Seed Co. v. Kalo Co., 333 US 127 [1948]
General Electric Co v De Forest Radio Co, 28 F 2d 641 [1928]
Genetic Systems Corporation v Roche Diagnostics GmbH & Dade Behring Marburg GmbH [2004] G0002/03
Gilead Sciences Pty Ltd v Idenix Pharmaceuticals LLC (2016) 117 IPR 252; [2016] FCA 169
Gottschalk V. Benson, 409 U.S. 63 [1972]
Grant v Commissioner of Patents [2006] FCAFC 120
Hartranft v Wiegmann, [1898] USSC 78
Harvard Oncomouse T19/90 [1990] OJ EPO
Harvard/Transgenic animal T-315/03 [2005] EPOR
IBM/Computer Product Program T 1173/97 [1998] EPOR
Illumina, Inc. v. Ariosa Diagnostics, Inc. 952 F.3d 1367 [2020]
In re Roslin Institute (Edinburgh), 750 F.3d 1333, 1334 Fed. Cir. [2014]
International Business Machines Corporation v Commissioner of Patents [1991] FCA 811
International Stem Cell Corporation [2016] APO 52
International Stem Cell Corporation v Comptroller General of Patents, Designs and Trade Marks [2014] C-364/13 at 39
InterPharma Pty Ltd v Hospira, Inc (No 3) [2017] FCA 1536
Joos v Commissioner of Patents [1972] HCA 38
Kingdom of the Netherlands v European Parliament and Council of the European Union (2001) C-3 77/98
Leland Stanford/Modified Animal [2002] EPOR
Mackay Radio & Telegraph Co. v. Radio Corp. of America, 306 U.S. 86 [1939]
Mayo Collaborative Services v. Prometheus Laboratories, Inc., 132 S.CT. 1289 [2012]
Meat & Livestock Australia Limited v Cargill, Inc [2018] FCA 51
Medi-Physics, Inc. G0001/07 [2010]
Merck & Co Inc v Olin Mathieson Chemical Co, [1958] USCA4 73; 253 F 2d 156, 164 [1958]
Monsanto Technology v Cefetra 428/08 [2010] ECR I-6765
National Research Development Corporation v Commissioner of Patents (1959) 102 CLR 252
Oliver Brüstle v Greenpeace eV C-34/10 [2011]
Parker v. Bergy, 563 F.2d 1031 (C.C.P.A. 1977) [*Bergy I*]
Parker v. Flook, 437 U.S. 584 [1978]
Plant Genetic Systems N.V., et al v Greenpeace Ltd. T356/93 [1995] OJ EPO 545
PPG Industries Ohio, Inc. [2004] G0001/03
Rapid Litig. Mgmt. v. CellzDirect, Inc., 827 F.3d 1042, 1049, 119 USPQ2d 1370, 1374 [2016]
Relaxin/Howard Florey Institute T 0272/95 [2002] EPOR
Rescare Ltd v Anaesthetic Supplies Pty Ltd [1992] FCA 811
Rozenberg and Co Pty Ltd v Velin-Pharma A/S [2017] APO 61
Sanofi-Aventis Australia Pty Ltd v Apotex Pty Ltd (No 3) [2011] FCA 846 240, 243
Sequenom, Inc. v Ariosa Diagnostics, Inc. [2019] FCA 1011
Shell Development Co. v. Watson, 149 F. Supp. 279, 280 [DC 1957]
Sun Pharmaceutical Industries (Australia) Pty Ltd v Tasmanian Alkaloids Pty Ltd [2018] APO 7

The Upjohn Company v Akzo Nobel Pharma B.V. T791/96 [1999]
TPI Enterprises Ltd v Tasmanian Alkaloids Pty Ltd [2018] APO 8
Vanda Pharmaceuticals Inc. v. West-Ward Pharmaceuticals, 887 F.3d 1117 [2018]
Viacom/Computer related invention T 0208/84 [1986] EPOR
Wisconsin Alumni Research Foundation G0002/06 [2008]
Wisconsin Alumni Research Foundation T1374/04 [2006]

Legislation/Directives

Directive 98/44/EC of the European Parliament and of the Council, on the legal protection of biotechnological inventions, *Parliamentary Debates*, 06 July 1998, 30)
European Patent Convention, opened for signature 5 October 1973, (entered in force 7 October 1977)
Leahy-Smith America Invents Act 2011 (US)
Patents Act 1990 (Cth)
The Constitution of the United States, created on 17 September 1787, ratified on 21 June, 1788

Treaties/Declarations/Regulations/Guidelines

Convention on the Unification of Certain Points of Substantive Law on Patents for Invention, opened for signature 27 November 1963, (entered in force 1 August 1980) [Strasbourg Convention]
European Convention on the International Classification of Patents for Inventions (Paris, December 19, 1954) ETS No. 17
European Convention relating to the Formalities required for Patent Applications (Paris, December 11, 1953) ETS No. 16
European Patent Office, Guidelines for Examination, List of exceptions, G-II 5.3
Intellectual Property India, 'Guidelines for Examination of Computer Related Inventions (CRIs)' 2017
IP Australia (2016), *Manual of Practice and Procedure*
Marrakesh Agreement Establishing the World Trade Organization, opened for signature 15 April 1994, 1867 UNTS 3 (entered into force 1 January 1995) annex 1C (*Agreement on Trade-Related Aspects of Intellectual Property Rights*) ('*TRIPS*')
Patent Cooperation Treaty (PCT), opened for signature 19 June 1970, (entered into force 24 January 1978) art 22
Recommendation 23 of the First Session of the Council of Europe (September 8, 1949, Doc 110)
United States Patent and Trademarks Office (2019), *Manual of Patent Examining Procedure*

Internet Sources

'A Bioprinting World Map' *3DPrint.com*, (Web page, 19 June 2019) https://3dprint.com/246670/a-bioprinting-world-map/
"Benefits Of Bioprinting - Brinter™ Bioprinter - 3D Bioprinting", *Brinter - 3D Bioprinting Solutions* (Webpage, 2022) https://www.brinter.com/benefits-of-bioprinting/
Clare Scott, University of California San Diego's 3D Printed Liver Tissue May Be the Closest We've Gotten to a Real Printed Liver, https://3dprint.com/118932/uc-san-diego-3d-printed-liver/

‘Commission Decision On Setting Up A Commission Expert Group On Development And Implications Of Patent Law In The Field Of Biotechnology And Genetic Engineering’, *Ec.Europa.Eu* (Webpage, 2012) https://ec.europa.eu/docsroom/documents/14893/attachments/1/translations

Derwent. 2021. *Derwent Innovation - Derwent*. [online] Available at: https://clarivate.com/derwent/solutions/derwent-innovation/

‘Engineered’, *Lexico powered by Oxford*, https://www.lexico.com/definition/engineered

Epo.org. 2018. *EPO - Validation of European patents in Cambodia (KH) with effect from 1 March 2018*. [online] Available at: https://www.epo.org/law-practice/legal-texts/official-journal/information-epo/archive/20180209.html

Examination Practice following the High Court decision in *D'Arcy v Myriad Genetics Inc*. https://www.ipaustralia.gov.au/about-us/news-and-community/news/examination-practice-following-high-court-decision-darcy-v-myriad

Hornick, John F and Kai Rajan, *The 3D Bioprinting Patent Landscape Takes Shape as IP Leaders Emerge* (July 7, 2016) 3D Printing Industry https://3dprintingindustry.com/news/3d-bioprinting-patent-landscape-takes-shape-ip-leaders-emerge-84541/

‘Human’, *Merriam-webster*, https://www.merriam-webster.com/dictionary/human; ‘Organism’, *Merriam-webster*, https://www.merriam-webster.com/dictionary/organism

IP Australia, *Patent Analytics on Stem Cell Technologies Underlying Regenerative Medicine*, Available at: https://www.ipaustralia.gov.au/tools-resources/publications-reports/patent-analytics-stem-cell-technologies-underlying-regenerative

IP Australia. 2022. *Types of patents*. [online] Available at: https://www.ipaustralia.gov.au/patents/understanding-patents/types-patents

Lens.org. 2021. *The Lens - Free & Open Patent and Scholarly Search*. [online] Available at: https://www.lens.org/lens/search/patent

List Of Sections Amended In 2022 Revision - Guidelines for Examination”, *Epo.Org* (Webpage, 2022) https://www.epo.org/law-practice/legal-texts/html/guidelines/e/m.htm

Patent quality, *USPTO*, https://www.uspto.gov/patents/patent-quality

PatSeer. 2021. *PatSeer*. [online] Available at: https://patseer.com/

Quinn, Gene et al, “LexisNexis IP - Trends in Patent Prosecution - Nov 30, 2021 - IPWatchdog.Com | Patents & Intellectual Property Law,” *IPWatchdog.com | Patents & Intellectual Property Law* (November 14, 2001) https://ipwatchdog.com/videos/lexisnexis-ip-trends-patent-prosecution-nov-30-2021/

Sakharkar, Ashwini, “3Dbio Implants 3D-Bioprinted Ear In A First-In-Human Clinical Trial”, *Inceptive Mind* (Webpage, 2022) https://www.inceptivemind.com/3dbio-implants-aurinovo-3d-bioprinted-ear-first-human-clinical-trial/25005/

‘State Of Play Of The Implementation Of Directive 98/44/EC’, *Ec.Europa.Eu* (Webpage, 2007) https://ec.europa.eu/docsroom/documents/14892/attachments/1/translations/en/renditions/pdf

Stemcell.ny.gov. 2021. *What is the difference between totipotent, pluripotent, and multipotent? | NYSTEM*. [online] Available at: https://stemcell.ny.gov/faqs/what-difference-between-totipotent-pluripotent-and-multipotent

Thevantagepoint.com. 2021. *The VantagePoint*. [online] Available at: https://www.thevantagepoint.com/

UK Intellectual Property Office, ‘Examination Guidelines for Patent Applications relating to Biotechnological Inventions in the Intellectual Property Office’ 2010. https://www.gov.uk/government/publications/examining-patent-applications-for-biotechnological-inventions

USPTO Report Shows Intellectual Property-Intensive Industries Contribute $5 Trillion, 40 Million Jobs to U.S. Economy, *USPTO*, https://www.uspto.gov/ip-policy/economic-research/publications/reports/uspto-report-shows-intellectual-property-intensive

GPSR Compliance

The European Union's (EU) General Product Safety Regulation (GPSR) is a set of rules that requires consumer products to be safe and our obligations to ensure this.

If you have any concerns about our products, you can contact us on ProductSafety@springernature.com

In case Publisher is established outside the EU, the EU authorized representative is:

Springer Nature Customer Service Center GmbH
Europaplatz 3
69115 Heidelberg, Germany

Batch number: 08362458

Printed by Printforce, the Netherlands